Science and F

Science and Fiction – A Springer Series

This collection of entertaining and thought-provoking books will appeal equally to science buffs, scientists and science-fiction fans. It was born out of the recognition that scientific discovery and the creation of plausible fictional scenarios are often two sides of the same coin. Each relies on an understanding of the way the world works, coupled with the imaginative ability to invent new or alternative explanations—and even other worlds. Authored by practicing scientists as well as writers of hard science fiction, these books explore and exploit the borderlands between accepted science and its fictional counterpart. Uncovering mutual influences, promoting fruitful interaction, narrating and analyzing fictional scenarios, together they serve as a reaction vessel for inspired new ideas in science, technology, and beyond.

Whether fiction, fact, or forever undecidable: the Springer Series "Science and Fiction" intends to go where no one has gone before!

Its largely non-technical books take several different approaches. Journey with their authors as they

- Indulge in science speculation—describing intriguing, plausible yet unproven ideas;
- Exploit science fiction for educational purposes and as a means of promoting critical thinking;
- Explore the interplay of science and science fiction—throughout the history of the genre and looking ahead;
- Delve into related topics including, but not limited to: science as a creative process, the limits of science, interplay of literature and knowledge;
- Tell fictional short stories built around well-defined scientific ideas, with a supplement summarizing the science underlying the plot.

Readers can look forward to a broad range of topics, as intriguing as they are important. Here just a few by way of illustration:

- Time travel, superluminal travel, wormholes, teleportation
- Extraterrestrial intelligence and alien civilizations
- Artificial intelligence, planetary brains, the universe as a computer, simulated worlds
- Non-anthropocentric viewpoints
- Synthetic biology, genetic engineering, developing nanotechnologies
- Eco/infrastructure/meteorite-impact disaster scenarios
- Future scenarios, transhumanism, posthumanism, intelligence explosion
- Virtual worlds, cyberspace dramas
- Consciousness and mind manipulation

More information about this series at http://www.springer.com/series/11657

Russell Blackford

Science Fiction and the Moral Imagination

Visions, Minds, Ethics

Russell Blackford
School of Humanities and Social Science
University of Newcastle
Callaghan, New South Wales
Australia

ISSN 2197-1188   ISSN 2197-1196   (electronic)
Science and Fiction
ISBN 978-3-319-61683-4   ISBN 978-3-319-61685-8   (eBook)
DOI 10.1007/978-3-319-61685-8

Library of Congress Control Number: 2017947530

Printed on acid-free paper

This Springer imprint is published by Springer Nature
The registered company is Springer International Publishing AG
The registered company address is: Gewerbestrasse 11, 6330 Cham, Switzerland

For Peter Nicholls

A Note to Readers

In what follows I sometimes use the letters "SF" for "science fiction" or "science fictional." This is mainly based on considerations of flow or euphony: on occasion, writing "SF" sounds less repetitive or awkward. If inconsistency about something like this bothers you, you have been warned.

I have avoided the abbreviation "sci-fi" even though this increasingly predominates, especially in mass media discourse. Many older aficionados of the science fiction genre still find something offensive about it. More important, perhaps, is my residual sense that "sci-fi" refers mainly to space opera, or at least to action-adventure stories with a science fictional twist of some kind. Thus, it works to refer to a movie like *Rogue One: A Stars Wars Story* as an example of sci-fi. The same might even apply to Ann Leckie's ambitious Imperial Radch series. But consider Mary Shelley's *Frankenstein*, George Orwell's *Nineteen Eighty-Four*, or Margaret Atwood's *The Handmaid's Tale*. These all possess the formal characteristics that define the mode or genre of science fiction. I have no compunction, therefore, about calling them SF novels, but to my ear there would be something very strange in referring to any of them as "sci-fi."

In the past, Atwood has been resistant to being considered a science fiction writer. The problem is that her work often does take the form of science fiction, even though it is not the sort of thing typically evoked when people talk about or use the expression "sci-fi." It has no extraterrestrial aliens or battles in outer space. But science fiction is a far broader phenomenon than space opera.

Each chapter of what follows provides a reference list of literary critical works (and similar) that I have cited. However, most of the creative works that I discuss are novels and movies, so I have not usually provided citation details beyond the name of the author, or director, and the date of first publication or cinematic release. This is usually sufficient for the reader to locate a copy. In some cases, I've thought it helpful to provide a bit more information in the text.

In most cases, I provide citation information, such as date of publication, only the first time a work is mentioned. As an appendix, however, I have included a list of works referred to more than in passing. This could also serve as a recommended reading/viewing list for somebody looking to develop an overall perspective on the science fiction genre. It is a slightly idiosyncratic list, not so much because I have odd personal tastes (as these things go) as because I'm approaching science fiction from a particular perspective. Still, it's no more idiosyncratic than other such lists.

Another warning—this book contains spoilers, lots of them. On one hand, I have not gone out of my way to reveal endings pointlessly. On the other hand, the interpretation of a novel, movie, or other work often depends on what happens at the end. In many cases, there was nothing for it but to describe endings and comment on what they imply.

Finally, *Science Fiction and the Moral Imagination* is intended to be accessible to almost anyone who is likely to pick it up (or to find it while browsing online). I assume some prior interest in the science fiction genre and some general interest in literature, but I don't assume deep knowledge of either. Nor do I use a lot of critical, philosophical, or scientific jargon. Despite its particular concern—the intersection of science fiction and moral philosophy—*Science Fiction and the Moral Imagination* could function reasonably as an introduction to SF for someone with only limited knowledge.

But if all has gone well, it also offers enough depth to enrich the understanding even of deeply read students of the genre. That, dear readers, is for you to judge.

Acknowledgments

I have dedicated this book to Peter Nicholls, the great encyclopedist of science fiction. For four decades, no one has engaged meaningfully with the genre without benefiting from his scholarly work.

Damien Broderick has been a mentor, colleague, and friend for over thirty years now. During that time, I've learned an enormous amount from him—about science fiction and about the difficult crafts of creative writing and literary criticism. I hope he will find some merit in this modest volume.

Another mentor was the late Norman Talbot, the supervisor of my first doctoral dissertation (the one in English literature), which I completed in the early 1980s. Justin Oakley supervised my second doctoral dissertation (this time in philosophy) more than two decades later. I learned much from both of them, and I've kept in mind their high scholarly standards.

Van Ikin and Sean McMullen were my coauthors of a previous book, *Strange Constellations: A History of Australian Science Fiction* (Greenwood Press, 1999). Both are fine scholars, and I benefited from working with them. The experience of researching and writing *Strange Constellations* hardened me for the rigors of another scholarly monograph relating to the SF genre. In addition, Van Ikin published much of my early criticism and scholarship— back in the 1980s and 1990s. He gave me the confidence that I can do this job. As editor of the journal *Science Fiction: A Review of Speculative Literature*, and in many other ways, he has made a crucial contribution to historical and critical understanding of the genre.

Thanks, too, to Gregory Benford, who has been kind to me in many ways and was instrumental in the publication of this book. And thanks to Angela Lahee, my supportive and patient commissioning editor at Springer.

My wife, Jenny Blackford, is a poet, author, scholar, IT expert, and extraordinarily sharp-eyed reader. She has added value to this project at many stages and in countless ways (the same applies to all my books). My heartfelt thanks to her, as ever.

Thanks to many other friends with whom I've discussed and debated science fiction and/or moral philosophy over very many years. I'll forget someone important if I try to name you all individually.

Finally, the title of this book is partly inspired by that of *Science Fiction: History, Science, Vision* by Robert Scholes and Eric S. Rabkin (Oxford University Press, 1977), a work that has influenced my general thinking about the SF genre since I first read it many years ago.

A book such as this discusses a large number of novels, short stories, plays, movies, and other cultural products. I've taken all possible care, including rereading many novels and stories that I'd last broached years, or even decades, ago. Still, there are bound to be some errors and misunderstandings. I'm entirely to blame for those—I hereby absolve everybody else mentioned in the paragraphs above.

Contents

1

Introduction: Science and the Rise of Science Fiction

Change and the Future

The future is another country. Human societies have always experienced change, whether from invasions and warfare, from mass religious movements, or from plague, famine, and natural disasters. But the revolutions in science and technology during the seventeenth, eighteenth, and nineteenth centuries brought something unprecedented. They greatly altered humanity's knowledge base, and with it they transformed previous understandings of humanity's place within the natural world. They changed the existing methods of industry and production, the arrangements for social organization, and, perhaps less obviously, the existing moral norms and varieties of cultural expression. As the Industrial Revolution unfolded, at first in Great Britain, but then in other European societies and their colonies, Western civilization experienced something altogether new: continual—and above all visible—change that was driven and shaped by advances in technoscience.

And so, humanity discovered the future. Previously, some religious and mythological systems had described grand cycles of time or an eventual (most likely supernatural) end of worldly things. These systems did not, however, speculate about future societies continually morphing in response to changes in the available knowledge base and technologies. As the scientist and social commentator J.D. Bernal observed much later in European modernity, human beings normally take accidental features of their own societies to be axiomatic features of the universe, likely to continue until supernaturally interrupted: "Until the last few centuries this inability to see the future except

© Springer International Publishing AG 2017

R. Blackford, *Science Fiction and the Moral Imagination*, Science and Fiction,
DOI 10.1007/978-3-319-61685-8_1

as a continuation of the present prevented any but mystical anticipations of it" (Bernal 1970 [1929], 12). But as Bernal elaborates, that assumption ceased to be tenable.

The distinguished science fiction author, critic, and scholar James Gunn makes a similar point: "Until the Industrial Revolution, to the average citizen the future did not exist in the sense we understand the word today" (2006, 209). As Gunn acknowledges, it was always known that there were days, years, and cycles of seasons to come, but any disruptions of these were likely to be bad news—the results of warfare, plague, and natural disasters. Humanity was less likely to look forward to a better time than to look back to an imagined Paradise or Golden Age.

Writers such as Bernal, Gunn, and the American literary theorist and critic Robert Scholes (1975, 11–21) rightly emphasize the transforming role of industrialization in the eighteenth and nineteenth centuries. But let us keep in mind the reconceptualization, commencing with the seventeenth-century Scientific Revolution, of humanity's imagined place in the universe. The rise, consolidation, and extension of science destroyed human exceptionalism.

What, then, is science? The Oxford English Dictionary defines a relevant usage of the word "science" that it traces to 1600:

A branch of study that deals with a connected body of demonstrated truths or with observed facts systematically classified and more or less comprehended by general laws, and incorporating trustworthy methods (now esp. those involving the scientific method and which incorporate falsifiable hypotheses) for the discovery of new truth in its own domain.

The word "scientist" is much more recent, dating to the 1830s and 1840s, when it was introduced as a general term—analogous to "artist"—to include practitioners of the specific sciences. The term was then applied retrospectively to historical individuals such as Galileo Galilei, Johannes Kepler, and Sir Isaac Newton, who saw themselves as natural philosophers, mathematicians, or students of specific areas of learning.

As historians and philosophers of science are painfully aware, it is difficult to draw a line between science and pseudoscience, between the sciences and the humanities, or between scientists and others who might be either legitimate scholars or mere pseudoscientists.[1] It is often assumed that scientists are

---

[1] Worse, there is a further distinction between the natural sciences and the social sciences. Throughout this book, I primarily have in mind the natural sciences whenever I refer to "science" or "scientists." That is, I think, the usual assumption in English. But contrast the German term *Wissenschaft*, which has a far broader meaning.

distinct because they employ a specific "scientific method," but that is doubtful. Science often looks for causal influences that are not directly observable, so it relies heavily on hypothetico-deductive reasoning: the formulation and testing of hypotheses. However, hypothetico-deductive reasoning is not restricted to the work of scientists, and indeed it is employed in all areas of scholarly inquiry and in much everyday problem solving.

There is much variation of approaches within the sciences as to how far they depend upon hypothetico-deductive reasoning and how far they emphasize other techniques, such as close, systematic observation, inductive reasoning, and efforts to isolate confounding influences. Nonetheless, we often do know scientists when we see them, even if they lived before the generic concept of a "scientist" was invented in the nineteenth century. How so?

During the seventeenth century, science increasingly began to investigate very small and very distant phenomena. In the centuries that followed, science turned, also, to the very long history of our world and its forms of life. The Scientific Revolution saw a heightened level of systematic and precise investigation of nature, using increasingly sophisticated techniques. Sciences such as physics and astronomy relied on advances in mathematics, controlled experiments, precisely engineered equipment, and new instruments of observation such as Galileo's telescope. Much of this did, indeed, add greatly to the *precision* and *power* of hypothetico-deductive reasoning. It is difficult to draw a clear line between science, as it is now understood in English-speaking countries, and rational inquiry more generally (including the humanities). However, there was a ferment of new methods in the seventeenth century, together with a sense of cooperation, rivalry, and excitement in the process of making discoveries and developing techniques. Any scholarly account of the rise of modern science (such as Gaukroger 2006) reveals a package of approaches that was historically innovative and dramatically effective in opening up the universe to human investigation.

As the sciences took shape, their practitioners were able to study phenomena that had previously resisted human efforts. These included very distant and vastly out-of-scale phenomena such as those studied by astronomers, very small phenomena such as the detailed composition and functioning of our bodies, and (somewhat later, with the advent of scientific geology) phenomena from deep in time before human artifacts, buildings, or written records. By the early decades of the nineteenth century, the emerging sciences were starting to imagine, and communicate, the extreme vastness of time as well as space.

The science of geology suggested that we live on the surface of an incomprehensibly old planet with a similarly incomprehensible number of years still to come. This idea of deep time has since been confirmed, elaborated, and expanded by scientists from numerous disciplines, and, all in all, a new understanding of the cosmos has emerged. Gunn suggests that Darwin's theory of evolution, first publicly revealed at the end of the 1850s, may have been more important to human understanding (and the development of science fiction) than the effects of innovations in technology. According to Gunn, biological evolution, including the Darwinian mechanism of natural selection, "suggested a different concept of humanity, not as special creation but as natural and evolutionary creatures" (2006, 30).

Even in the twenty-first century, our scientific picture of the cosmos in space and time—and of ourselves—remains incomplete, but much of it is now too well-evidenced for any reversal to be imagined. The scientific picture radically challenges older concepts of human exceptionalism, while offering marvels more astonishing, if sometimes less humanly intuitive, than anything found in pre-scientific mythmaking.

The revolutions in science and technology during the centuries of European modernity introduced new ideas about the universe, ourselves, and the future:

- We inhabit an incomprehensibly vast universe whose origins lie deep in time. Our own beginnings as a species are temporally remote, and our final destiny is unknown.
- We are, ourselves, the results of natural processes, much like other living things. From the new perspective, human exceptionalism is no longer tenable.
- All known social and cultural forms, and specifically those we have experienced in our individual lifetimes, are significantly mutable. Even the relatively near future may turn out very strange by the standards of those now living.

At least one other historical process tended to confirm all this. During approximately the period encompassed by the scientific and industrial revolutions, though beginning a century or so earlier, exploration and colonialism brought the cultures of Europe into contact with what seemed like strange—sometimes hostile—environments and peoples. For some European intellectuals, this provoked a sense of the historical contingency and precariousness of existing cultures and civilizations. The practices and beliefs of particular human cultures increasingly appeared at least somewhat arbitrary—somewhat the products of chance—having co-evolved to fit with each other while being shaped by local economic and other material circumstances. By the nineteenth century, this cohered well with the undeniable evidence that traditional forms of economic and social organization were vulnerable to technological transformation.

The past few centuries, going back at least to Galileo's great discoveries in the early years of the 1600s, have seen a series of breakthroughs in conceptualizing the universe, ourselves, and the human future. All of this offered new opportunities for storytelling.

Science Fiction and the Wellsian Imagination

Imaginative writers of a new genre, known by the 1930s as science fiction, responded—whether with enthusiasm, anxiety, or regret—to the phenomena of modern science, rapid technological change, and industrialization.

As early as the seventeenth century, some works of fiction incorporated the nascent scientific understanding of the universe and humanity's place in it. Johannes Kepler's *Somnium, Sive Astronomia Lunaris* (completed c. 1608–1609, but not formally published until 1634) is sometimes claimed to be the first science fiction novel, but it has none of the characteristics that we normally associate with novels, such as telling a complex story and including characters with at least some appearance of psychological plausibility. It also has little in the way of the characteristics of science fiction (which we'll soon come to). In particular, there is no attempt to develop a story that carries a sense of verisimilitude, given certain unusual assumptions. *Somnium* is a geography of the Moon's surface, based on the best observations that had been made prior to astronomical use of telescopes. These are introduced via a thinly developed fictional narrative.

In introducing *Somnium* to modern readers, John Lear suggests that Kepler's major purpose was to criticize human exceptionalism, "to demonstrate to his fellows that the human animal was not the central figure of the cosmos, that the heavens did not wait upon his home planet, earth" (1965, 66). Perhaps this overstates the case. Apart from setting out what was known of the lunar landscape, *Somnium* was evidently intended as an allegory about the ability of science to obtain knowledge once it frees itself of igorance. Nonetheless, it is a narrative composed at the very dawn of the Scientific Revolution, and it has an unmistakably scientific theme. It suggests that things might not be as they appear to us from our vantage point on Earth. Many assumptions that come naturally to us because of our location in the universe might seem dubious—and might have seemingly strange equivalents—if we lived elsewhere.

Somnium can be considered a work of proto-science fiction, rather than a fully fledged science fiction novel. Yet it foreshadows themes that SF writers explored in the nineteenth century and beyond. On the one hand, there is the trust that science can obtain knowledge of kinds that previously eluded human efforts. At the same time, there is the sense of a physically greater cosmos than was previously imagined. Along with this goes a recognition of our relative smallness in the total scheme, and of our limited understanding.

The sense of human smallness is well conveyed in much of the SF canon. One example is H.G. Wells's *The War of the Worlds* (1897), one of the most influential of the author's many novels. Here, the narrator sees the invading Martians, with their great war machines, as standing to us as we stand to ants. In the end, the Martians' efforts are not defeated by human soldiers, guns, and warships, despite some heroic efforts against the odds, but by the invaders' susceptibility to our planet's micro-organisms. Before their conquest can proceed far, they sicken and die, leaving it unknown whether Martians might one day make another attempt. Though the Martian invasion fails, it reveals humanity's puniness and vulnerability within the immensity of the larger cosmos, alerts our species to the unknown dangers or benefits that could arrive unexpectedly from space, opens up imaginative possibilities, and promotes a sense of common humanity. Or so the narrator concludes. Whatever the future might bring, and despite the terrible cost in human life, the overall effect of the invasion turns out to be salutary. The invasion has driven home a more scientific picture of humanity's place in the universe—one that was, of course, available in its essentials at the time when Wells was writing in the 1890s, but also one that is not easily grasped even now.

In his scholarly book *The Future as Nightmare: H.G. Wells and the Anti-utopians*, Mark R. Hillegas discusses the influence of Wells and (especially) the twentieth-century reaction against his more utopian novels. Hillegas suggests that the Wellsian imagination is drawn to certain subjects:

> It is fascinated by the revelations of man's place in space and time given to us by science, fascinated by the vistas of astronomy, particularly the death of the world and the vastness of interstellar space, fascinated by the vision of geological epochs, the evolution of life, and the early history of man vouchsafed by geology, paleontology, and archaeology. (Hillegas 1967, 15)

The Wellsian imagination is fascinated, that is, by the vast cosmos opened up for our inspection by science—and by the knowledge that we are relatively small, new, and fragile within this larger context.

Arthur C. Clarke's *Rendezvous with Rama* (1973) is one of many science fiction novels that exemplify such a vision and such an imagination. It is set in the year 2130, by which time our Solar System includes several main inhabited worlds: Earth, Luna, Mars, Mercury (whose inhabitants are referred to as Hermians), Ganymede, Titan, and Triton. As background, we learn that much of northern Italy, including Venice, Padua, and Verona, was destroyed in 2077 by a huge meteor. This led to the establishment of Project SPACEGUARD to ensure that nothing similar could happen again. Early in the narrative, SPACEGUARD detects a huge alien artifact, which is given the name "Rama" by astronomers. Leaders on Earth and the colonized worlds are keenly sensitive to the possibility of another disaster from space, so a crewed spaceship, the *Endeavour*, is sent to study the strange object.

The main events involve exploration of Rama's interior. Commander William Norton and his crew have limited time to investigate the artifact, make some sense of it, and establish whether it is dangerous. They are dwarfed by the artificial world that they discover, one with hurricanes, electrical storms, its own sea, high cliffs, and giant waves. There are also bizarre partly robotic, partly biological creatures with a kind of ecology of their own. Rama's environment continually tests the astronauts' ingenuity, and the artifact itself defies human knowledge and understanding to the very end.

Rama uses a space drive that no one understands, and we never see its masters. Humanity's encounter with Rama demonstrates not only the immensity of the universe—and hence our own relative smallness—but also what seems almost like Rama's contempt for human actions and defiance of our efforts at understanding. Rama simply enters the Solar System, then leaves again, making no effort to contact us. After tapping matter from the Sun to refuel—the description of this is precise, yet glorious—it heads off in the direction of the Greater Magellanic Cloud, beyond the Milky Way, leaving no clues behind. As Peter Brigg expresses it, "The very fact that Rama has paid man no attention whatsoever during his brief invasion is a stunning reminder of man's tiny place in the scale of the universe" (1977, 44).

What Is This Thing Called Science Fiction?

Notwithstanding some early works such as Kepler's *Somnium*, science fiction is best seen as a child of the nineteenth century. It recognizes the deep, rapid, intra-generational change in modern societies, and this was not possible until the Industrial Revolution was well underway.

Isaac Asimov has stated that "Science fiction can be defined as that branch of literature which deals with the reaction of human beings to changes in science and technology" (1981, 82). This is not adequate as a *definition* of science fiction, since SF is surely not the *only* branch of literature that responds in such a way. Much mainstream literature that no one would classify as science fiction responds to changes in science and technology—and perhaps deals thematically with its effects on individuals and societies. Even modern fantasy responds, if indirectly, to industrialization, and Asimov himself points out that J.R.R Tolkien's great work of modern fantasy, *The Lord of the Rings* (originally published in three volumes, 1954–1955), can be read as a lament over the industrialization of life and the landscape in England and other affected countries. At one level, as Asimov tells us, "The Mordor of *The Lord of the Rings* is the industrial world which is slowly developing and taking over the whole planet, consuming it, poisoning it" (1981, 293). This element of nostalgia for an earlier, more pastoral way of life is clearly present in Tolkien's work, and it can be seen from the beginning of modern heroic fantasy in the late nineteenth century, when this genre was invented by authors such as William Morris. Nonetheless, Asimov is correct to link science fiction to the revolutions in science, technology, and industrial production that began in the seventeenth and eighteenth centuries, and then accelerated through the nineteenth, leading up to the present day.

Asimov also describes science fiction stories as dealing with events that could not have taken place against current or past social backgrounds, as understood at the time of writing, but could take place against a background that might, conceivably, "be derived from our own by appropriate changes in the level of science and technology" (1981, 18). He interprets the idea of conceivable scientific advances rather loosely, so as to include unlikely innovations such as time travel and travel faster than the speed of light, and he then makes a crucial point:

> Given this definition of science fiction, we can see that the field can scarcely have existed in its true sense until the time came when the concept of social change through alterations in the level of science and technology had evolved in the first place. (Asimov 1981, 18)

This is a useful insight, and it explains why nothing like a recognizable SF genre became established until the nineteenth century. Indeed, it was not labeled as a specific genre until the early decades of the twentieth century. Tinkering slightly with Asimov's account, we can understand science fiction as a literary response—not the only one—to humanity's discovery of the future.

As Asimov emphasizes, science fiction was not possible for as long as society was viewed as essentially static (barring events such as war and imagined events such as supernatural intervention). Accordingly, it arose in response to the Industrial Revolution, which first became visible during the second half of the eighteenth century. Asimov concurs with Brian W. Aldiss's well-known assessment of Mary Shelley's *Frankenstein; or, The Modern Prometheus* (1818) as the first genuine science fiction novel,[2] and he considers its timing, early in the nineteenth century, unsurprising (Asimov 1981, 19; compare Aldiss 1973, 7–39, or Aldiss and Wingrove 1986, 36–46). This seems, I submit, as plausible a starting point as any. It represents a beginning for science fiction exactly two centuries ago.

During the nineteenth century, SF writers increasingly speculated about technological devices not yet invented, conceived of diverse possible futures, or filled the immense universe, the imaginary future, and the deep past with exotic locations for tales of adventure. As the pace of social and technological change accelerated during the twentieth century, narratives of technological innovation and humanity's future prospects became even more culturally prominent. But despite science fiction's familiarity in contemporary culture, its boundaries remain contentious. Strangely, perhaps, no definition of the genre commands universal scholarly assent.

Fortunately, certain formal elements do feature in most definitions. To summarize these, I'll employ the terms "novelty," "rationality," and "realism," though I need to explain the special senses in which I use these words. First, science fiction is marked by what the theorist Darko Suvin describes as an element of "strange newness" or "a *novum*" (1979, 4). As Suvin explains the point, the *novum* is any violation of, or break with, the empirical environment of the author's own society and historically recorded societies. It might be the presence of new technology, for example, or it might be a new scientific discovery. Alternatively, the narrative might take place within a future society.

Second, a science fiction narrative has a rational element. Here, I have in mind a concept similar to Suvin's when he states that whatever is novel in an SF narrative is "perceived as *not impossible* within the cognitive (cosmological and anthropological) norms of the writer's epoch" (1979, viii). Rationality in this sense involves a degree of deference to science.

Third, there is an element of realism. This does not require a strong commitment to believable character motivations. However, science fiction narratives have at least a basic level of verisimilude and literalness. That is,

---

[2] Aldiss may not have been the first to make this claim about *Frankenstein*, but his authority certainly lent it great credibility.

the events depicted are to be understood by the audience as actually happening within the imagined universe of the story. They are meant literally, even if they have a further allegorical or metaphorical level of meaning.

The element of novelty in SF narratives is somewhat constrained by those of rationality and realism. Science fiction presents situations and events that are strange and new, but they are not supernatural. To the extent that supernatural entities and influences do enter into what otherwise looks like an SF narrative, its status as pure science fiction is at least somewhat undermined. Such narratives are more like a hybrid—admittedly, a rather common one—of science fiction and fantasy.

Suvin's approach has the advantage of explaining how a narrative could qualify as science fiction even though a very similar (or even identical) text might not have qualified if produced in a different era with different standards of what is novel and what is considered, in Suvin's words, "not impossible." For Suvin, what matters—perhaps separating science fiction from other forms of writing such as myth and fantasy—is that what is portrayed is considered within the bounds of possibility at the time and place of a narrative's composition.

This brings me to an important theoretical point: it is more strictly accurate to describe science fiction as a narrative *mode* rather than, more vaguely, as a narrative *genre*. Genres are usually defined by their settings (think of Westerns) or their conventional storylines (think of detective stories and romances). In referring to narrative modes, I follow Northrop Frye's conception of the idea in his *Anatomy of Criticism* (1957), a classic and astonishingly erudite overview of literature and its critical study. We need not accept all the details of Frye's theory of modes, but it can be useful to categorize narratives in accordance with the main characters' powers of action (Frye 1957, 33). Are these characters, for example, gods or godlike, possessed of superhuman abilities, superior in power to us (though still human), much like us in power, or inferior in power to us (Frye 1957, 33–34)?

Once we think in this way, we can see science fiction as defined by the sorts of events that can take place and, more specifically, the sorts of actions that the characters can perform. In brief, they can perform an indefinitely wide range of actions—wider than those available to ordinary people within the empirical environment of the author's own society, for example—with the caveat that their actions must be, in a loose sense, scientifically or empirically possible. And so, within Frye's conception of a narrative mode, we can speak and write accurately of a science fiction mode. In the pages that follow, I'll normally avoid that language and write contentedly enough of a "science fiction genre," as almost everybody else does. This is convenient for me and probably more intuitive for readers. However, it is worth keeping in mind that what makes science fiction distinctive is the expanded, yet bounded, range of things that can happen and things that its characters can *do*.

To this point, I have largely followed Suvin's approach to identifying science fiction's formal characteristics, though I suspect that Suvin would be more willing than I am to dismiss some narratives as not science fictional at all if their premises seem too fantastic. To be clear, however, he mirrors Asimov in taking a lenient view of what counts as scientifically possible for the purpose of defining science fiction's boundaries. Suvin suggests that we understand science as "an open-ended corpus of knowledge," and so a science fiction story might rely on imagined scientific discoveries of the future that "do not contravene the philosophical basis of the scientific method in the author's times" (Suvin 1979, 68). For example, he understands his approach as including Ursula K. Le Guin's *The Dispossessed: An Ambiguous Utopia* (1974), which is partly about future developments in fundamental physics.

In one way, Suvin might even be *too* lenient. Taken literally, his approach could lead us to classify works from prescientific eras as science fiction. For example, the various gods, monsters, and supernatural events of Homer's *Odyssey* were not part of the experienced, empirical reality of the time but were not considered impossible within the prevailing worldview. Should we classify *The Odyssey* as an early work of science fiction? I think that would be a mistake. Permit me, then, to emphasize that SF's element of rationality should be understood as a degree of deference to the *scientific* understandings of the time—if any such understandings existed. Putting this another way, science fiction normally pays at least some lip service to a *scientific view of the world*. Whatever wonders it presents are at least handwaved as having a scientific explanation.

To complicate matters further, many stories that are marketed, read, and discussed as science fiction fit somewhat uncomfortably into a narrative genre that characteristically focuses on future innovations in technology and on technologically-transformed future societies. Consider the following categories:

- Alternative (or "alternate") history stories, such as Philip K. Dick's *The Man in the High Castle* (1962) or Michael Chabon's *The Yiddish Policemen's Union* (2007). The former portrays a world in which the Axis powers won World War II. In the latter, an alternative history of the war involved the creation of a Jewish refuge in Alaska and the eventual destruction of Berlin with nuclear weapons.

- Stories that portray contact with alien civilizations, even if set in the present and not involving any technological innovations on the part of human beings.
- Stories that portray certain kinds of natural disasters, such as asteroid impacts.
- Stories that portray the aftermath of global war, whether or not any new military technologies are imagined.
- Stories that portray humanity's prehistoric past, such as Jean M. Auel's *The Clan of the Cave Bear* (1980).

Some of these seem marginal to the science fiction genre, and whether they are regarded as SF in individual cases might depend on circumstances—not least on whether the author is generally regarded as a science fiction writer. Nonetheless, these kinds of narrative combine novelty, rationality, and realism to at least some extent. Their composition and intelligibility depend, moreover, on modern scientific knowledge and/or a general sense of human societies' historical contingency. Nuclear holocaust stories would be impossible but for the scientific advances that led to the development of nuclear weapons. Stories of asteroid impacts or mankind's remote past were not readily imaginable until science began to describe the vastness of space and time, and to theorize about human origins.

Stories of alien contact are central to the SF genre, but some other categories, such as prehistory stories and alternative history stories, might be thought of as peripheral categories of science fiction with a family resemblance to the more central kinds. (This need not detract from the quality and importance of some individual works that fall in one of the relevant categories. Dick's *The Man in the High Castle* is a good example.)

A thornier set of issues arises when science fiction is compared to the modern fantasy genre. Fantasy is notable for its depiction of gods, demons, and other supernatural beings, scientifically inexplicable events, and/or the effective operation of magic in some form. Its plots and character types tend to draw on those found in myth, legend, fairytale, and romance. On initial consideration, all of this contrasts sharply with science fiction's rationality, realism, and concern with technological change. And yet, SF and fantasy share similar marketing and an extensively overlapping audience. What should we make of this?

Rather than drawing a bright line, we should acknowledge that the contrast between fantasy and science fiction is less than absolute. There are similarities, as well as differences. The work of modern fantasy writers—such as William Morris at the end of the nineteenth century and J.R.R. Tolkien in the twentieth—can resemble that of some SF writers in its portrayal of imaginary societies, and sometimes entire worlds, although fantasy usually (there are exceptions) does not locate these in the future or on faraway planets. Fantasy narratives can also display a kind of rationality: the operation of magical forces, powers, and methods may be internally consistent within a fantasy narrative, giving a sense of problems being solved within a set of "rules" even if they are not claimed to have any scientific justification.

Conversely, science fiction's deference to real science can be token or minimal, and the fictional science it describes might then be little more than magic by another name. Almost any fictional event can be given *some* kind of scientific, or at least pseudoscientific, rationalization if the "science" concerned is that of an imagined time or place with an extraordinarily advanced understanding of the natural order. As a result, much of what has been marketed as science fiction can be understood equally well as a variety of modern fantasy ornamented with technological trappings.

In the upshot, scientific ideas of time, space, and other worlds are easily co-opted for the depiction of adventures and conflicts in exotic locations. It is, then, explicable that many narratives blend science fiction and fantasy elements, that adventure-oriented SF and fantasy frequently blur into each other, and that the SF and fantasy genres often exert a similar appeal. Suvin would, I think, deplore this and insist that many supposedly SF narratives are fantasy in disguise. Perhaps so, but the considerable blurring and hybridization of "F&SF" is a cultural reality for scholars to cope with as best they can.

Nonetheless, with all appropriate caveats stated and complications acknowledged, science fiction is a genre or (better) a mode of narrative that is characterized by the elements of novelty, rationality, and realism. It typically depicts future developments in social organization, science, and/or technology, and its main thematic focus is on the effects of technological change (whether upon individuals or societies). This is an idealized description, because science fiction is extremely varied, and it has been shaped by historical contingencies and commercial realities.

Among those realities, some popular fiction is not marketed as science fiction even though it clearly shares SF's formal characteristics. This is especially so of technothrillers by such authors as the late Michael Crichton. Crichton's *Jurassic Park* (1990), for example, could easily have been badged more clearly as a science fiction novel if that had been commercially justified. It is science fiction for my purposes, as are the movies adapted from it, beginning with *Jurassic Park* (dir. Steven Spielberg, 1993). The practical reality is that the marketing of novels, stories, movies, and so on as science fiction or as something else—for example as thrillers, or even as "literary" works—often depends more on promoting a brand than on any formal characteristics of the narratives themselves.

Science Fiction Meets the Moral Imagination

As I've mentioned, science fiction is often employed to portray adventures and conflicts in exotic locales. When its tropes are used more seriously, however, it often explores the social and psychological effects—and hence the moral significance—of scientific and technological innovations. With its greatly extended narrative possibilities, science fiction can illuminate the social impact of change, propose blueprints for a better future, or implicitly criticize any naive optimism about where the human species is headed. Storytellers working in the SF field have used its tropes to challenge traditional moral norms—and suggest alternatives—and to dramatize moral dilemmas in creative ways. Recurring themes in science fiction include the design and functioning of future societies, terraforming and cosmic engineering, reshaping ourselves with technology, and questions about our treatment of non-human persons (perhaps extraterrestrial aliens or advanced artificial intellects). Science fiction writers have employed the genre's tropes to engage with a wide variety of moral questions.

In what follows, then, I'll select exemplary texts and productions from the science fiction genre in order to demonstrate typical themes and issues. The result should appeal to readers who wish to enhance their understanding of moral philosophy through literature and popular culture. My primary aim, however, is not to use SF stories to illustrate philosophical ideas. There is nothing objectionable about this technique: it can be pedagogically useful, and the results can sometimes be enjoyable, and even educational, all round. But it is not my main purpose in writing this book.

For a comprehensive introduction to modern philosophy, using science fiction stories as a basis for discussion, readers might wish to consult *Philosophy Through Science Fiction* by Nichols, Smith, and Miller (2009). This offers some philosophical conclusions that I would approach with caution—what philosophical book doesn't?—but it is a useful work of pedagogy. As its title makes clear, it is an introduction to philosophy rather than a philosophically oriented introduction to science fiction. Susan Schneider's edited volume *Philosophy and Science Fiction* (2009) is useful for much the same purpose. It concentrates on metaphysical, rather than moral, questions, and uses SF stories to illustrate them.

Science Fiction and Philosophy (it's OK in refs)

Compared with the Nichols, Smith, and Miller volume, in particular, my range is narrower in one sense. For the purposes of this book, I am interested in moral questions—and not, for example, in philosophical questions relating to free will, the existence of God, or the nature of time. Moreover, I place a strong emphasis on appreciating the science fiction genre through examining its contributions to humanity's moral imagination. I propose to examine SF's engagement with moral questions mainly for its own sake, rather than as a springboard to teach philosophy. I approach the exercise from the viewpoint of a scholar with training in both philosophy and literary scholarship. I hope to contribute to our understanding of how the genre fits together historically and thematically.

The following chapters may, then, be useful for philosophy teachers—I hope they will be!—but unlike some other books *Science Fiction and the Moral Imagination* is not designed as a tool for direct use in teaching philosophy.

A Roadmap

In Chapter 2, I will provide a somewhat comprehensive history of the science fiction genre. I claim no great originality for anything in the chapter. My intention at this point is not to be controversial, or even especially creative, but to provide an integrated account that is sufficiently orthodox, accurate, and detailed for readers to get their bearings. The chapter will help place novels, short stories, and other works of science fiction in their artistic and historical contexts.

Chapter 3 begins with a brief introduction to moral philosophy—at least sufficient for some orientation in that respect. The purpose of the chapter is to show that science fiction's tropes can function as enabling devices: they enable SF authors to engage with philosophical questions, including some of those that arise in moral philosophy. I offer some varied examples.

In Chapter 4, I am interested in the moral norms that might prevail (for better or worse) in humanity's future. When characters and events are distanced from us in space or time, one effect is to open up the possibility of moral alternatives. As I explain and illustrate, this applies to institutions and conduct related to sexuality, families, and other issues that are heavily moralized in existing societies.

Chapter 5 concerns a key science fiction theme: the acceptable uses of science and technology, given the great power that these can grant. Among other issues, I discuss science fiction's typical images of scientists and, more generally, of science fiction heroes who employ science and technology. This provokes reflections on the Promethean, Galilean, or Frankensteinian aspects of many significant characters from the SF canon.

In Chapter 6, I look more closely at non-human or mutated persons. These may be alien beings from different planets or times, artificial beings such as robots, cyborgs, and androids, or strange humans of one kind or another (often resulting from genetic mutation). All of these can raise questions about what it means to be human. They also provoke us to ask how we should treat those who are not like ourselves, yet have characteristics that seem to call for our respect and consideration. Science fiction raises questions about how we should treat human Others and how we would treat the possible more-than-human or non-human Intelligent Others of the future.

Chapters 6 and 7 are closely related—indeed, I have made some uncomfortably arbitrary decisions about which works to discuss in which chapter. That said, Chapter 7 focuses specifically on the use of technology to enhance human capacities. It concerns human beings who have obtained greater-than-human powers of action as a result of intelligent interventions. These might sometimes be a curse more than a blessing. The idea of transforming ourselves to expand human capacities receives approbation in the SF genre at least as early as Wells—for example in *The World Set Free* (1914)—but it also meets with much resistance.

In Chapter 8, I explore the question of great responsibility: the large-scale events, projects, and consequences that feature in many works of science fiction. What does SF teach us about great responsibility? More worryingly, do SF's scale and spectacle undermine its claim to seriousness? In this chapter, I sum up and draw some lessons, touching briefly on science fiction's relationship to a broadly posthumanist ethic.

Concluding Remarks

In what follows, I will concentrate mainly on written science fiction, but this does not reflect any particular disdain for other media, such as film. James Gunn has suggested that the only SF movies that qualify as both good films and good science fiction are *Things to Come* (dir. William Cameron Menzies, 1936) and *2001: A Space Odyssey* (dir. Stanley Kubrick, 1968) (Gunn 2006, 123). Gunn has much to say in praise of *Things to Come*, in particular. At one point, he emphasizes the power of its final sequences and he refers to similar reactions from other SF writers:

Raymond Massey's final statement of man's destiny still sounds the clear, pure call of mainstream science fiction. I have been surprised at the number of science fiction writers of my generation, such as Isaac Asimov and Fred Pohl, who expressed the same reaction to this film. (Gunn 2006, 155)

Gunn is not the only critic who laments the track record of science fiction cinema. In his influential 1995 monograph, *Reading by Starlight*, Damien Broderick mentions that he avoids discussing SF cinema because it "is starved fare in terms of the rich gravies of cognitive estrangement available in print form" (1995, 111). Paul J. Nahin takes a similar approach in his recent book on science fiction and religion, deciding to mention movies only in passing, and television programs not at all, because of what he sees as generally low standards (2014, xiv).

Overall, these critics have a point. Although Broderick's book was published more than two decades ago, his complaint is still largely justified, in that many SF movies are formulaic. Even when they are based on strong material, they can ruin it through dumbing down and sentimentalizing. Nonetheless, there are many longstanding masterpieces of SF cinema—certainly more than Gunn acknowledges—and some tightly directed, thematically sharp movies have been produced just in the last few years.

My focus on written science fiction is not the result of disdain for SF in other media, but mainly to save space. A much longer book would be required if I were to venture far into discussing science fiction in cinema, radio, television, comics, and computer games (the last being a medium in which I claim no expertise in any event). To keep this book manageable, I've decided to discuss movies where they provide good examples to make a point, but I will not give them the same depth of coverage that I'll give SF in the form of prose narratives. I'll neglect SF in other media almost entirely—with mentions only in my historical discussion of the genre or when an irresistibly salient example comes to mind.

Also to avoid writing a much longer book, I will concentrate almost entirely on works originally written in English. To be honest, this also reflects my own limitations. There is now a vast body of SF composed in languages other than English, including but not limited to French, German, Spanish, Portuguese, Russian, Polish, Chinese, and Japanese. From what familiarity I have with this body of work—such of it as has been translated into English—I'm sure much of it deserves study and commentary. That, however, is beyond my present expertise. I'll confine my discussion accordingly, with the main exceptions being a few works of great historical importance. Unfortunately, this will leave my treatment of some authors seriously underdone relative to their importance within the genre.[3]

Some authors currently writing in languages other than English are producing especially important work. One whose work I do discuss in the following chapters is the Chinese maestro Cixin Liu (Liu Cixin), whose fiction has only recently begun to appear in English translation. At the present time, he is one of the world's leading SF authors by any reasonable standard, a worthy successor to past grandmasters, such as Isaac Asimov and Robert A. Heinlein. Liu seems to be steeped in Asimov's oeuvre in particular.

Within the limits I've set out, I will range fairly widely, commenting on a variety of narratives, although I will focus whenever convenient on historically influential novels and movies. To convey an idea of the current state of the genre, I will also give weight to some quite recent work. My selection of narratives for discussion will doubtless show biases, but I will not attempt to draw strong conclusions from a small selection of works contrived for the purpose. Of course, my topic is a large one, and there is an ocean of relevant material from which I have had to select. In many cases, no doubt, I could have chosen different texts for discussion, though some choices were almost inevitable: among them, *Frankenstein*, H.G. Wells's great scientific romances of the 1890s and early 1900s, Le Guin's *The Dispossessed*, and the Culture series by Iain M. Banks. In any event, what follows will offer a framework to consider texts that did not make my particular cut.

Science fiction has a rich, proud history. Without more scene-setting, let us turn our minds to it.

References

Aldiss, B. W. (1973). *Billion year spree: The history of science fiction*. London: Weidenfeld & Nicolson.

Aldiss, B. W., & Wingrove, D. (1986). *Trillion year spree: The history of science fiction*. London: Gollancz.

Asimov, I. (1981). *Asimov on science fiction*. Garden City, NY: Doubleday.

Bernal, J. D. (1970). *The world, the flesh, and the devil: An inquiry into the future of the three enemies of the rational soul*. London: Jonathan Cape (Orig. pub. 1929).

---

[3] Poland's Stanislaw Lem is one author who immediately comes to mind as deserving more commentary than I've given.

Brigg, P. (1977). Three styles of Arthur C. Clarke: The projector, the wit, and the mystic. In J. D. Olander & M. H. Greenberg (Eds.), *Arthur C. Clarke* (pp. 15–51). New York: Taplinger.

Broderick, D. (1995). *Reading by starlight: Postmodern science fiction*. New York: Routledge.

Frye, N. (1957). *Anatomy of criticism: Four essays*. Princeton, NJ: Princeton University Press.

Gaukroger, S. (2006). *The emergence of a scientific culture: Science and the shaping of modernity, 1210–1685*. Oxford and New York: Oxford University Press.

Gunn, J. (2006). *Inside science fiction* (2nd ed.). Lanham, MD: Scarecrow.

Hillegas, M. R. (1967). *The future as nightmare: H.G. Wells and the anti-utopians*. New York: Oxford University Press.

Lear, J. (1965). *Kepler's dream*. Berkeley and Los Angeles: University of California Press.

Nahin, P. J. (2014). *Holy sci-fi!: Where science fiction and religion intersect*. New York: Springer.

Nichols, R., Smith, N. D., & Miller, F. (2009). *Philosophy through science fiction: A coursebook with readings*. New York: Routledge.

Oxford English Dictionary. Third edition (online version). (November 3, 2015.)

Schneider, S. (Ed.). (2009). *Science fiction and philosophy: From time travel to super-intelligence*. Hoboken, NJ: Wiley-Blackwell.

Scholes, R. (1975). *Structural fabulation: An essay on fiction of the future*. Notre Dame, IN: Notre Dame University Press.

Suvin, D. (1979). *Metamorphoses of science fiction: On the poetics and history of a literary genre*. New Haven, CT: Yale University Press.

2

Science Fiction: A Short History of a Literary Genre

Emergence

Science fiction's boundaries are porous, and it is difficult to state exactly when the genre emerged. Some works written in the early phases of the Scientific Revolution can be seen as prototypes, among them Johannes Kepler's *Somnium* and Francis Bacon's *The New Atlantis* (1629). The latter is a utopian fragment glorifying science and technology. Some early narratives of voyages and discoveries show elements comparable to those of later science fiction, though few of them much resemble the genuine article. Mark R. Hillegas regards the "Voyage to Laputa" part of Jonathan Swift's *Gulliver's Travels* (1726)[1] as the only significant science fictional voyage narrative in the eighteenth century (Hillegas 1967, 9–10), but even this is a fantastical satire of the science of the time rather than an SF narrative in its own right.

As it took a more recognizable form during the nineteenth century, the science fiction genre responded to the phenomena of science, technology, and industrialization. The Industrial Revolution ushered in a new concept of change and the future, with its numerous technological innovations—beginning, as James Gunn observes, with "steam and its impact upon manufacturing and transportation" (2006, 30), but then involving other novelties, including electricity, medical advances, and the automobile. Brian W. Aldiss, a distinguished British author and historian of the genre, has put a persuasive case for Mary Shelley's *Frankenstein*, first published in 1818, as the first true SF

---

[1] The full, original title to Swift's work is the ponderous *Travels into Several Remote Nations of the World. In Four Parts. By Lemuel Gulliver, First a Surgeon, and then a Captain of several Ships.*

© Springer International Publishing AG 2017
R. Blackford, *Science Fiction and the Moral Imagination*, Science and Fiction,
DOI 10.1007/978-3-319-61685-8_2

novel (Aldiss 1973, 7–39; Aldiss and Wingrove 1986, 25–52). As I suggested in Chapter 1 this seems as plausible a starting point as any. Shelley famously depicts Victor Frankenstein's use of scientifically based technology to create something entirely new in the world: a powerful, but unfortunately repulsive, artificial man.

Shelley later wrote *The Last Man* (1826), which is of interest for its inclusion of characters based on the personalities of Lord Byron and Percy Bysshe Shelley. This novel is a melodramatic story of love, jealousy, war, and political intrigue, all set at the end of the twenty-first century. The events it portrays culminate in the destruction of humanity by an incurable plague that arrives in the British Isles via Turkey and continental Europe. Lionel Verney—who actually does seem to be the last man left on Earth—writes the story in the first person, addressing his words to the dead or to whoever might one day see them. (At one point, Verney contemplates that an Edenic couple might have been spared somewhere to repopulate the world.) While *The Last Man* is open to interpretation, it is better seen as a lament for the failure of Romanticism's revolutionary potential than as a meditation on the human future.

It is reasonable, nonetheless, to consider *The Last Man* an early science fiction novel, given its depiction of an altered future world and its imagination of a purely secular end to humanity. In previous centuries, humanity's eventual end—if any such thing were even contemplated—was invariably assigned a supernatural cause. That said, the world-destroying epidemic in *The Last Man* is a more catastrophic version of the plagues that had swept across Europe in previous centuries and in the author's own lifetime. Furthermore, as the novel's action commences the world has not greatly changed. England has established a republican system of government, but the social organization depicted appears much like that of the early nineteenth century. Likewise, the technological and economic base of society is unchanged: there is little or nothing in the methods of transport, communication, housing, or warfare that would be unfamiliar to Shelley's contemporaries.

Some of Edgar Allan Poe's stories from the 1830s and 1840s have science fiction elements, and James Gunn regards Poe's "Mellonta Tauta" (1848) as possibly the first true story of the future (Gunn 2006, 10). Unlike *The Last Man*, it does portray a future society with unfamiliar ideas and practices. The story is set in the year 2848—thus, one thousand years after its date of composition—and its Greek title can be translated as "future things" or "things of the future" (or perhaps even as "things to come"). This sounds rather portentous, but the narrative takes the form of one person's rambling, gossiping, speculation-filled letter to a friend. In fact, it is more like a series of diary entries, beginning on April 1—April Fools' Day, of course—and it is composed by a well-educated but deeply misinformed individual, Pundita, who reveals that she is on a pleasure excursion aboard a balloon.

In Poe's version of the future, humanity has explored the Moon and made contact with its diminutive people. However, much knowledge from the nineteenth century has become garbled and (at least) half lost. "Mellonta Tauta" thus sheds doubt on historians' confident interpretations of the practices of other peoples living in earlier times. For example, Pundita holds forth about the rival philosophical systems of "Aries Tottle" (obviously Aristotle) and "Hog" (evidently a conflation of Francis Bacon and the eighteenth-century Scottish author James Hogg).

Despite her considerable ignorance of what she is talking about, Pundita seems to some extent a spokesperson for Poe in criticizing various social, political, and philosophical follies, as he evidently sees them, of his own time. Amidst her other musings, Pundita marvels at the foolishness of the ancient system of democratic elections—unfortunately misunderstanding them, but not so far beyond recognition as to take away the humor. She also engages in archeological speculation about the strange ways of the long-ago "Amriccans," who were, she thinks, undoubtedly cannibals. Some of the satire is aimed at one of Poe's greatest contemporaries, the British philosopher and statesman John Stuart Mill, and at the moral theory of utilitarianism that Mill championed throughout his career.

If Pundita is Poe's spokesperson to any extent at all, Poe is unimpressed not only by Mill but by any kind of philosophy grounded solely in logical analysis and empirical observation. This is, however, all unclear since Pundita is shown as far from reliable and Poe was no mystic. Poe elaborated a system for understanding the universe, and our place within it, in his philosophical and spiritual treatise *Eureka: A Prose Poem* (1848). This includes much of the material published in "Mellonta Tauta," but readers of both texts face difficulties when attempting to construe their point and tone. As for "Mellonta Tauta," much of the humor will be lost on modern readers who are not steeped in the same cultural referents as its original audience. Nonetheless, Poe laid a foundation for the development of satirical SF set in future, greatly altered societies.

A more substantial body of work that resembles modern science fiction emerged around 1860, particularly with the stories and novels of the French author Jules Verne. Verne is best known for novels in which highly advanced (for the time) science and technology enable remarkable journeys, as in *Five Weeks in a Balloon* (1863), *Journey to the Centre of the Earth* (1864), *From the Earth to the Moon* (1865), and *Twenty Thousand Leagues Under the Sea* (first serialized 1869–1870). H.G. Wells's career as a writer of what were then known as scientific romances commenced a few decades later, with a group of short stories that led up to his short novel *The Time Machine* (1895). The importance of this work for later SF writers cannot be overstated. Darko Suvin writes, without hyperbole, that "all subsequent significant SF can be said to have sprung from Wells's *The Time Machine*" (1979, 221). Wells followed up *The Time Machine* with his first full-length "scientific romance," *The Island of Doctor Moreau* (1896).

During this same period, the American author Edward Bellamy published his enormously successful novel *Looking Backward, 2000–1887* (1888). In this utopian work, Julian West, the protagonist and narrator, falls asleep in 1887 from the actions of a mesmerizer. He finally wakes in the year 2000, discovering himself in a better world that has been brought about in the interim with no violence. Instead, it has arisen from a process of corporate growth until each of the political entities in the world, including the USA, has become a single corporate organization entirely merged with the state. *Looking Backward* is overtly didactic, although Bellamy's style is sufficiently lively and transparent to keep us reading, and there is, perhaps, just enough suspense for us to care throughout about what might happen next. Most of the text consists of explanations by the kindly Dr. Leete (in whose home West awakens) of the industrial and economic functioning of American society in 2000. Dr. Leete explains at length—and sometimes with insufferable smugness—how things are now done in better ways than in the late nineteenth century.

Although *Looking Backward* takes the form of a novel, it is essentially a tract setting out the blueprint for a radically socialist or communist utopia. One arresting feature (at least for a moral philosopher) is its acknowledgment that people will have different moral intuitions under different social circumstances, though some intuitions might be considered more pure, less distorted, than others. Hence, all West's instincts about moral issues are shown as distorted by the narrow-minded and ungenerous society in which he was raised.

During the early years of his literary career, Bellamy was a prolific author. However his reputation now rests entirely on *Looking Backward*. Then again, the importance and influence of this book in its time—and its indirect influence ever since—is incalculable. It prompted a political movement (albeit a short-lived one), has been read by millions of people, and inspired responses from major authors such as William Morris and, of course, Wells.

If *Looking Backward* provided a template for utopian novels set in the future, Wells provided the first template for futuristic dystopian fiction with *When the Sleeper Wakes* (first published in serial form in 1898–1899). He later revised this for book publication as *The Sleeper Awakes* (1910). In both versions, the novel tells of a man who falls into a kind of trance and wakes up over two hundred years later in a very altered London. As it turns out, he was bequeathed investments while he slept. His trustees managed these astutely and ruthlessly, and were spectacularly successful as a result. By the time the eponymous Sleeper wakes, his trustees—the White Council—have established themselves as plutocratic world masters. The dystopian nature of this future world becomes increasingly clear, as the Sleeper (along with Wells's readers) learns that more than a third of its people live in wretchedness, effectively enslaved by the Labour Company. The latter is owned by the Sleeper himself, and is a descendant of the Salvation Army, which his trustees acquired during his long sleep.

In the late nineteenth and early twentieth centuries, SF elements appeared in many utopias, lost-world novels, and stories of near-future geopolitical disruption. Adventure novels of the time, such as H. Rider Haggard's classic story of a lost world, *She: A History of Adventure* (first serialized 1886–1887), frequently took place in remote, exotic, and often mildly erotic, locations. The use of interplanetary settings took this a step further. The first published novel by Edgar Rice Burroughs, *A Princess of Mars* (originally published in serial form in 1912), epitomized the trend. Planetary romance of the kind favored by Burroughs defined one pole of early science fiction, emphasizing action and adventure in an alien setting.

Another approach was the near-future political thriller. Works of this sort, most notably *The Battle of Dorking: Reminiscences of a Volunteer* by George Tomkyns Chesney (1871), were a prominent component of the literary scene in the late nineteenth and early twentieth centuries. They portrayed future wars and invasions, typically involving racial conflict. Their plot points did not necessarily include any new technologies or methods of warfare, and they often contained melodramatic and blatantly racist elements. Nonetheless, they were intended as serious speculations about the vulnerability of established nations and societies to militarism and demographic pressures.

All of these forms of early science fiction have continued to the present day. Literary scientific romances, particularly inspired by those of Wells and those who reacted to him, have maintained a pedigree partly independent of what I will call "genre science fiction" and define in the following section. Post-Wellsian scientific romance includes the work of Olaf Stapledon, as well as that of Aldous Huxley, George Orwell (Eric Arthur Blair), and many others. Canadian author Margaret Atwood and the American author Richard Powers might be seen as current exponents. Some influential literary works in the early decades of the tradition took the form of plays or play sequences, as with George Bernard Shaw's sequence *Back to Methuselah (A Metabiological*

Pentateuch) (first published 1921; first performed 1922), and the plays of the great Czech author Karel Čapek. The tradition of scientific romance has produced some striking dystopian—or anti-utopian—narratives, as with E.M. Forster's "The Machine Stops" (1909), Huxley's *Brave New World* (1932), Orwell's *Nineteen Eighty-Four* (1949), and Atwood's *The Handmaid's Tale* (1985).

The Rise of Genre Science Fiction

Genre science fiction—by which I mean work produced for a distinctive "science fiction market"—dates from the 1920s and the 1930s, when SF was first conceptualized as a distinct branch of narrative literature and its direction was shaped by two great American editors: Luxembourg-born Hugo Gernsback and John W. Campbell, Jr. Neither invented science fiction, or claimed to have done so, and Edgar Rice Burroughs was already popular with his tales of exotic adventure, sometimes set beyond Earth. Burroughs was successful throughout the eras of Gernsback and Campbell and beyond, independently of the specialized SF magazines. Likewise, Philip Wylie carved out an extraordinary career that merits more attention from scholars. His 1930 novel, *Gladiator*, was published in book form by Alfred A. Knopf. *When Worlds Collide*, co-authored by Wylie with Edwin Balmer, was serialized in the generalist *Blue Book* magazine from 1932–1933.

However, Gernsback brought a fresh perspective. He particularly looked for stories that emphasized futurism, science, and gadgetry. As an editor, he wanted his writers to introduce young readers to the wonders of technology that would come to exist in scientifically advanced societies of the future. He first showed this sort of interest with the magazine *Modern Electrics* (1908) and his own prophetic, but risibly melodramatic, novel *Ralph 124C 41+* (first serialized 1911–1912). In the latter, a great twenty-seventh century inventor uses his technoscientific mastery to save a damsel from a series of situations of distress. Gernsback did not devote an entire magazine to his interest in fictionalized technoscience until 1926, when he launched *Amazing Stories*, often regarded as the first specialized science fiction magazine.

During the Gernsback era, the sub-genre now known as "space opera" became an established form. This type of narrative involves action on a galactic scale, or beyond, rather than mere adventures on the local planets of our solar system as in *A Princess of Mars* and its kind. Space opera often includes space-voyaging analogues to Earth-bound naval fleets. Its typical elements are contact with alien species, interstellar war, descriptions of immensely destructive weapons, and various kinds of super-science. One influential contribution to the science fiction genre from *Amazing Stories* was the serialized publication, in 1928, of E.E. "Doc" Smith's novel *The Skylark of Space*, which decisively established the sub-genre of space opera.

During the Gernsback era, Verne and Wells came to be thought of in retrospect as science fiction writers. Their work was often reprinted in specialist magazines alongside new stories that applauded the advance of science and technology (while often showing advanced technoscience causing dangers and crises requiring human resourcefulness). Gernsback initially coined the term "scientifiction," rather than "science fiction," but he adopted the latter usage for the first issue of *Science Wonder Stories* in 1929, and it became the established term in 1938 when it was incorporated into the title of John W. Campbell's *Astounding Science-Fiction* (Clareson 1990, 16).

Science fiction's so-called Golden Age, closely associated with Campbell, began in the late 1930s and lasted until the end of the 1940s. In 1937, Campbell began to assume editorial duties at the magazine *Astounding Stories* (originally founded in 1930 as *Astounding Stories of Super-Science*). In early 1938, he changed the magazine's name to *Astounding Science-Fiction* (the hyphen in "Science-Fiction" was later dropped; a couple of name changes later, the magazine is now *Analog Science Fiction and Fact*). For a short time, Campbell did little to alter *Astounding*'s direction, but by 1939 the magazine and the science fiction genre as a whole were being steered by his strong personality. Campbell published some of the early short stories of the distinguished British SF writer Arthur C. Clarke, though Clarke was not among Campbell's regular stable of contributors during the Golden Age. The Campbell stable did include Isaac Asimov, Robert A. Heinlein, L. Ron Hubbard, and A.E. van Vogt.

Some of these authors have fared better than others in respect of their longer-term reputations. Hubbard is now best known as the founder of Scientology, and van Vogt's reputation has, to say the least, not held up as well as Asimov's or Heinlein's (or Clarke's). The other major author to emerge in 1940s genre science fiction was Ray Bradbury, whose more poetic style did not mesh with Campbell's approach. He began his career in less regarded SF magazines than *Astounding* but ultimately built a reputation in the field comparable to those of Asimov, Heinlein, and Clarke. Arguably he achieved even greater fame in the wider literary world.

During the Golden Age, Asimov, Heinlein and others developed the concept of a future history of humanity, an idea first formalized by Heinlein, but expanded by Asimov. By the end of the 1940s, genre science fiction writers had created an inter-textual mythos involving the development and fluctuations of future galactic empires, with our planet as an ancestral home not necessarily remembered (see Clareson 1990, 31–33; Gunn 2006, 34–35). Beyond this shared understanding of future history, science fiction developed a network of familiar, easily-evoked icons, well understood by dedicated readers. Damien Broderick discusses the existence of an "extensive generic mega-text" in the field of science fiction—a concept that he develops from the work of literary theorist Christine Brooke-Rose—built up by numerous texts adopting and alluding to each other's ideas over many decades (Broderick 1995, 57–63). These shared ideas include, most obviously, monsters, spaceships, robots, and futuristic cities. Broderick suggests that the mega-text can expand, and that it has indeed been transformed by the introduction of more tropes and icons to the mix:

> When novelties like hyperspace and cyberspace, memex and AI (Artificial Intelligence), nanotech and plug-in personality agents are very quickly taken up as the common property of a number of independent stories and authors, we have the beginnings of a new mega-text. (Broderick 1995, 59)

As Broderick duly notes, none of these items possesses an inherent, unchanging meaning. For example, robots have been interpreted and reinterpreted multiple times, whether as dangerous, unsettling, helpful, comical, or in numerous other ways. "Yet," Broderick adds, "all these variants bear certain family resemblances, and tend to cohere about a limited number of narrative vectors" (1995, 60).

Isaac Asimov: Science Fiction as Problem Solving

Isaac Asimov's approach provides a useful case study, since it epitomizes Golden Age science fiction's emphasis on problem solving. Cumulatively, Asimov's body of fiction creates a sense of historical crises as puzzles to be resolved through logic and insight.

This can be seen in his Foundation Trilogy, one of the towering canonical texts of twentieth-century SF. Its three volumes—*Foundation* (1951), *Foundation and Empire* (1952), and *Second Foundation* (1953)—repackaged a series of stories that Asimov had originally published from 1942 to 1950, based around the "Seldon Plan." As the tale unfolds, thousands of years in the future, Hari Seldon, founder of the new science of psychohistory, establishes two foundations, said to be at the opposite ends of the Galaxy. He intends the two foundations to reduce a 30,000-year interregnum between the predicted fall of the Galactic Empire and the rise of a new empire. Using his expertise in psychohistory, he has predicted that this long interregnum can be reduced to "only" a thousand years, saving untold chaos and suffering. In the Foundation Trilogy, psychohistory is depicted as a highly-formalized method for understanding, predicting and (to some extent) controlling the direction of historical events.

Foundation traces the early history of the First Foundation as it deals with a series of crises. At each crisis point, the key to resolving the situation is an insight that is achieved by the Foundation's leaders—and was achieved in advance by Seldon—but is withheld from the reader until the end. This pattern changes somewhat in *Foundation and Empire*. The first part of the book deals with an attempted war against the Foundation by the Empire's last strong general, but the second introduces the Mule, a powerful mutant renegade who can control the emotions of others, even at a distance, and who begins to carve out his own empire. The Mule is such a freak of nature that the psychohistorical equations have not taken his abilities into account, and thus the course of history diverges from Seldon's plan. In *Second Foundation*, however, he is finally outguessed and outplayed by the Second Foundation's First Speaker. (The Speakers have similar powers to the Mule, though not on the same level.) As depicted by Asimov, the Mule is a truly formidable opponent who only ever suffers two defeats, neither of which is fatal to him or his empire. However, these setbacks give the Second Foundation the breathing space it needs to survive and restore the Seldon Plan.

After the Mule's death, there is a war between the remnants of his empire and the First Foundation, with the Foundation eventually prevailing. It is ultimately revealed, after much clever fencing around the issue, that the Second Foundation is actually based on Trantor, the decayed capital planet of the Galactic Empire, and that its current First Speaker is a seemingly innocuous farmer, Preem Palver. This more or less ties everything up, in terms of various mysteries, but we have penetrated only a few hundred years into the interregnum, leaving hundreds, or potentially thousands, of years to see how the final fate of the Galaxy will unfold.

Decades later, Asimov returned to the setting of the Foundation Trilogy with *Foundation's Edge* (1982), which jumps 120 years into the future from the events depicted at the end of *Second Foundation*. The problem is now that the Seldon Plan is unfolding so perfectly that it seems that there must be some outside interference. Some minds of the First Foundation conclude that the Second Foundation still exists, while some of the Second Foundation's minds conclude that an unknown agency is directing events. The book nears its end with a three-way standoff, as Asimov sets a tableau where the rival characters appear more or less equally matched. As usual, what matters is who has the deepest understanding of the situation. (In this case, the deadlock involves the planet Gaia, which is an emergent mind of enormous power, and mutually antagonistic representatives of the two Foundations.)

Asimov's other great achievement during the Golden Age was the development of one of science fiction's most celebrated and recurrent icons, the mechanical human being, or robot. Though he did not introduce robots into the SF mega-text, he employed the idea in an extraordinarily influential way. He devised his famous three laws of robotics to govern his robots' behavior: these laws required robots to preserve human life, obey human commands, and preserve themselves—in that order of priority—and the possible loopholes inherent in the three laws of robotics enabled Asimov to write many ingenious stories. These included two "robot detective" novels in the mid-1950s: *The Caves of Steel* (1954) and *The Naked Sun* (1957).

In *Foundation's Edge*, Asimov began to unite his Foundation mythos with his robot stories, and he continued this program in further novels published in the 1980s and 1990s: *The Robots of Dawn* (1983), *Robots and Empire* (1985), *Foundation and Earth* (1986), *Prelude to Foundation* (1988), and *Forward the Foundation* (published in 1993, after Asimov's death the previous year). Of all these, *Foundation and Earth* takes the story furthest into the future. In these late novels, Asimov reinterprets the storyline of the Foundation Trilogy within a much broader and more complex framework. However, the key in these novels is still that crises are successfully resolved by obtaining a deep understanding of the situation and acting laterally in accordance with it.

Another example of Asimov's approach is *The End of Eternity* (1955), which can be interpreted as a prequel to the Foundation Trilogy, although that knowledge is unnecessary to understand and enjoy the story. In this novel, we are shown a future world without space travel, let alone space colonization, although (in fact, *because*) the theoretical basis for *time* travel was discovered in the twenty-fourth century. Thereafter the required mathematics developed to a point where time travel became practicable, and a powerful organization, Eternity, was eventually established to utilize it wisely. Asimov shows us how Eternity interferes across the whole course of human history, extending far into the future.

Like the much later *Foundation's Edge*, *The End of Eternity* concludes with a three-way contest. In this case, the participants are the main character (a brilliant but somewhat narcissistic and paranoid operative for Eternity), his supervisor and mentor Laban Twissell (who attempts to preserve Eternity and to ensure that it was established on cue), and a woman from the very far future with her own plans for past human beings and for Eternity itself. She has the best understanding of the overall situation—and so she succeeds.

After the Golden Age

By the late 1940s, genre science fiction was changing. The destruction of Hiroshima and Nagasaki and the abrupt surrender of Japan turned the public's attention to science, and hence to science fiction, and more SF narratives began to appear outside of the specialist magazines, whether in the more prestigious slick magazines or in book form. Though the power of the atomic bomb forced the public to take science more seriously, it also contributed to a growing pessimism about the directions science was taking. Over the following decades, this fed back into the content of science fiction itself. Genre SF had shown a fundamental optimism under Campbell's lead, but pessimistic notes now entered.

From the late 1940s, new specialist magazines challenged Campbell's pre-eminence as an editor. In 1949, *The Magazine of Fantasy* appeared, edited by Anthony Boucher and J. Francis McComas. With its second issue in 1950, this became *The Magazine of Fantasy and Science Fiction* (Clareson 1990, 49). In 1950, the first issue of *Galaxy Science Fiction* appeared. In the late 1940s and throughout the 1950s book publication became increasingly important to the field, and all this introduced a greater variety of themes and attitudes. Isaac Asimov has suggested another element in the genre's increasing pessimism about science during the 1950s: writers increasingly entered the science fiction field who might previously have written other kinds of fiction. SF attracted them as other markets dried up. According to Asimov, this "meant that many of the new generation of science fiction writers had no knowledge of science, no sympathy for it—and were in fact rather hostile to it" (1981, 163).

During the 1950s, such authors as Alfred Bester, James Blish, Ray Bradbury, Damon Knight, Walter M. Miller, Frederik Pohl and Cyril M. Kornbluth, Clifford Simak, Theodore Sturgeon, and Kurt Vonnegut experimented with form, style, and tone. Bester's most successful work, such as *The Demolished Man* (1953) and *The Stars My Destination* (serialized 1956–1957; initially published in book form in 1957 as *Tiger! Tiger!*) shows "obsessed men driven to, if not beyond the borders of sanity" (Clareson 1990, 71). The team of Pohl and Kornbluth published with *Galaxy* during the 1950s. Their satirical *Gravy Train* was serialized in 1952, and revised in book form as *The Space Merchants* in 1953.

Ray Bradbury and Kurt Vonnegut, in particular, now brought science fiction to a wider audience than it had previously enjoyed (Clareson 1990, 49–50). Bradbury is probably best known for *The Martian Chronicles* (1950; published in the UK in slightly different form as *The Silver Locusts* in 1951) and *Fahrenheit 451* (1953). The linked stories that constitute *The Martian Chronicles* portray the colonization of Mars by humans. The first explorers inadvertently wipe out the ancient Martian civilization, since chickenpox is deadly to the Martians. Over the next few years, a frontier culture develops on Mars, until a long-feared nuclear war takes place on Earth in 2005—at which point almost all the colonists desert Mars and return to Earth.

Fahrenheit 451 depicts an anti-intellectual future America, where houses are fireproof and "firemen" burn books. As in many dystopian novels, the focus is on a well-positioned character—in this case one of the book-burning firemen—who loses trust in the system and begins a private rebellion. The satirical intent, directed at Bradbury's own society, needs no laboring. The novel's title, which refers to the temperature at which paper burns, has now become widely known as a way to refer to book-burning and suppression of ideas.

In many ways, however, as Thomas Clareson elaborates (1990, 115–116), the 1950s belonged to Heinlein. His first novel to appear in book form, *Rocket Ship Galileo*, was published in 1947. It was followed by a string of highly successful novels—the "Heinlein juveniles"—aimed at what would now be called the Young Adult market. Their success, along with reprinting of much of Heinlein's work from the magazines, made Heinlein a dominant figure.

The 1960s saw further experimentation. Some established writers adapted their approach to suit the times. Heinlein produced *Stranger in a Strange Land* (1961), perhaps the best known of all his novels, which celebrated sex and the body, advocated open sexual relationships, and satirized politics, organized religion, and traditional social *mores*. *Stranger in a Strange Land* appealed to an audience far beyond genre SF's usual reach—as did Frank Herbert's *Dune* (1965) and the 1960s novels of Kurt Vonnegut such as *Cat's Cradle* (1963) and *Slaughterhouse-Five* (1969), which reacted to the political assassinations of the Kennedys and Martin Luther King.[2]

Dune was originally published in *Analog* as two serialized stories, forming one unusually long book by the standards of the time for science fiction novels. It is set in the distant future on a faraway planet, Arrakis, with a complex social and political structure. Arrakis is a desert planet whose only non-human life

---

[2] *Slaughterhouse-Five* is ambiguously science fictional. It achieved great commercial and critical success, establishing Vonnegut for the first time as a major figure in the literary mainstream.

forms are gigantic worms that burrow in its sands. Against that backdrop, Herbert portrays complex political and religious struggles involving its main character, who becomes a messianic prophet. *Dune* was followed by numerous sequels, and its emphasis on world building and the planetary environment has since been profoundly influential.

During the 1960s and early 1970s, Heinlein wrote several more books, most notably *The Moon is a Harsh Mistress* (1966), which tells the story of a lunar colony's rebellion against Earth, analogous to the American Revolution. This is sometimes ranked as Heinlein's best novel, and it is undoubtedly one of the masterworks of the genre. However, his later work is often criticized as lacking shape and discipline, and there was a seven-year gap in his output from 1973 to 1980, when he was plagued by a series of health crises. Heinlein returned in the 1980s, writing several novels, of which *Friday* (1982) and the satirical *Job: A Comedy of Justice* (1984) showed him still near his best form.

In the early 1960s, under the editorship of E.J. Carnell and then the more aggressive Michael Moorcock (who took over in 1963), the leading British science fiction magazine, *New Worlds*, began to promote the New Wave style of science fiction. This largely rejected the genre's tradition of narrative realism, and instead purported to explore the mind's "inner space," a concept that was associated with J.G. Ballard in particular. The writing of the New Wave was rich in symbolism, surreal or inexplicable events, and allusions to compelling personalities such as Marilyn Monroe. It challenged sexual and other taboos, dealt with such subjects as environmental disaster, and often depicted Western civilization, or our entire planet, as doomed.

Unfortunately, it is unclear whose writing did or did not fall within the New Wave. Brian W. Aldiss is often considered a leading exponent, along with Ballard, although Aldiss disliked the term "New Wave." John Brunner is sometimes mentioned, but his relationship to the New Wave was ambiguous at best, and his outstanding successes in 1960s and 1970s were four powerful dystopian novels in which he experimented with an approach and style of his own: *Stand on Zanzibar* (1968), *The Jagged Orbit* (1969), *The Sheep Look Up* (1972), and *Shockwave Rider* (1975). His literary model, especially for *Stand on Zanzibar*, was the work of John Dos Passos.

In brief, the New Wave was originally associated with *New Worlds*. Its most obvious exemplar was Ballard and its leading advocates were Moorcock and the American writer and editor Judith Merril. However, the expression was soon used to cover broader changes in the style of British and American science fiction during the 1960s. In the US, Harlan Ellison published his blockbuster anthologies *Dangerous Visions* (1967) and *Again Dangerous Visions* (1972), for which he commissioned stories that might be regarded as controversial and subversive. These included British as well as American writers—among the contributors to *Dangerous Visions* were Aldiss, Ballard, and Brunner.

From this point, the science fiction genre displays multiple facets, and I can mention only a few. SF writers such as Philip José Farmer, Samuel R. Delany, and John Varley increasingly explored sexuality in their work. The prolific Philip K. Dick—whose stories and novels have frequently been adapted for the cinema—produced much of his best writing during the 1960s, including *The Man in the High Castle* and *Do Androids Dream of Electric Sheep?* (1968).

During the mid-1960s and early 1970s, another immensely prolific author, Robert Silverberg, produced some of his most studied and insightful work, including the award-winning novella "Nightwings" (1969), *A Time of Changes* (1971), and *Dying Inside* (1972). *A Time of Changes* depicts one man's rebellion against an oppressive culture that rejects individuality as an obscenity. *Dying Inside* is a sensitive depiction of declining powers—but in this case it is focused on a middle-aged man whose telepathic abilities are failing him. Joe Haldeman appeared as a powerful new talent with *The Forever War* (1974), a gritty anti-war novel that drew on the author's experience as a soldier in Vietnam. Golden Age values were kept alive in the work of Larry Niven and Jerry Pournelle, most notably in Niven's *Ringworld* (1970) and in the Niven/Pournelle collaboration *The Mote in God's Eye* (1974). Frederik Pohl, with *Gateway* (1977), published what many consider his best solo novel, and this is a prominent example of a veteran writer adopting some of the New Wave techniques (Harris-Fain 2005, 69).

During the late 1960s and the 1970s, there was also a vigorous interest in feminist and racial themes, sexuality, political theories, and utopian visions. Key works of the period were Ursula K. Le Guin's *The Left Hand of Darkness* (1969) and *The Dispossessed*, Joanna Russ's *The Female Man* (1975), Marge Piercy's *Woman on the Edge of Time* (1976), Samuel R. Delany's *Dhalgren* (1975), and Octavia E. Butler's *Kindred* (1979).

Of these, the huge, enigmatic, sexually explicit *Dhalgren* (1975) was another book that burst out of the confines of SF marketing and found a cult audience. Delany followed up with *Trouble on Triton: An Ambiguous Heterotopia* (originally published in 1976 as *Triton*, though with the same sub-title). This novel somewhat resembles the typical dystopian narrative in centering on a character who rejects the order of things and engages in a personal rebellion. There is, however, a twist. The society of freedom and economic plenty that Delany portrays is genuinely utopian or close to it. The protagonist is unable to find happiness because of his own shortcomings. In *Kindred*, Butler uses time travel as an enabling device to examine the American experience of slavery. *Kindred* does not attempt to rationalize time travel in any scientific way, but focuses on its main character's experiences when she is repeatedly—and unaccountably—hurled back in time to the late eighteenth or early nineteenth centuries. The novel has achieved recognition far beyond the SF field for its detailed and uncompromising portrayal of slavery in action.[3]

Writing under the pseudonym "James Tiptree Jr." (and for a few years as "Raccoona Sheldon") Alice B. Sheldon produced an extraordinary body of work from her first published story in 1968 until she died in 1987. Her pseudonym was exposed by a sequence of events surrounding her mother's death in 1977. Prior to this, Sheldon had been a focal point for debate because of the combination of her masculine pseudonym (it was widely known that "James Tiptree Jr." was a fictional name) and the feminist themes that often surfaced in her short stories of the time. While some readers guessed that she might be female, some high-profile commentators insisted that her work was recognizably masculine in style. Merely by writing as she did, while protecting her identity, Sheldon raised questions about whether any style of writing is ineluctably masculine or feminine.

Darren Harris-Fain observes that the SF field was fragmenting in the late 1970s and early 1980s (2005, 98–100). By this time, it was possible for a reader to focus on a particular type of science fiction, according to taste, but many writers and publishers were looking for something to reunify the field. Alas, this never happened. However, the cyberpunk movement of the 1980s synthesized many elements of what came before—and it left a durable legacy.

Cyberpunk and Beyond

The cyberpunk movement was influenced by the New Wave and by the politically engaged narratives of Samuel R. Delany and Joanna Russ, but it also incorporated elements of more traditional, Campbell-style science fiction. Other influences came from literary fiction with SF elements. For example, Thomas Pynchon's body of work, most notably *Gravity's Rainbow* (1973), profoundly influenced cyberpunk writers of the 1980s such as William Gibson. Another influence was William S. Burroughs' bizarre oeuvre—notably *The Naked Lunch* (1959), *The Soft Machine* (1961), *The Ticket that Exploded* (1962), and *Nova Express* (1966)—which portrays mind control, dystopian forms of social oppression, alien invasion, and transmutations to and from human form.

---

[3] For further discussion of *The Left Hand of Darkness*, *The Dispossessed*, *The Female Man*, and *Woman on the Edge of Time*, see Chapter 4.

Quintessential cyberpunk consists of fast-paced stories set in a near-future world where the power of national governments has been supplanted by that of wealthy corporations. These, in turn, are likely to be challenged by sub-cultures, cults, and criminal organizations. Cyberpunk writers depicted direct interfacing between human minds and advanced computers, events in computer-constructed virtual realities, and the activities of powerful artificial intelligences. All of this provided the "cyber" aspect of their writing, with blurring lines between reality and virtuality, and between human and machine. The "punk" aspect involved portrayals of street life, youthful rebellion, tough-guy attitudes and dress codes, and certain specific images, such as chrome, reflective glass, and architectural ruins.

The cyberpunk style was showcased in Gibson's early short stories, collected in *Burning Chrome* (1986), and in his first novel, *Neuromancer* (1984), which Gibson soon expanded into a trilogy.[4] Gibson's fiction of the 1980s is often characterized as dystopian, but this is simplistic. In *Neuromancer* and its sequels, the future is undeniably harsh; however, humanity has survived, individuals can still exert agency, and many aspects of what we are shown are actually pleasant in their way.

Neuromancer, in particular, has been much celebrated and equally imitated. High-profile cyberpunk writers, such as Gibson, Pat Cadigan, Bruce Sterling, and (later) Neal Stephenson, had a stunning impact on the SF field: their vision and sensibility have been absorbed into the ever-evolving SF mega-text, which means that the field as a whole now has a "post-cyberpunk" quality, apparent across all the relevant media. In the twenty-first century, science fiction's tropes and iconography emphasize machine intelligence and posthuman minds and bodies. Themes of alienation and rebellion abound, and with them images of ruined, desolate cityscapes. Even space opera set far from present-day life is now pervaded by images of darkness and alienation, and more generally by cyberpunk tropes.

In an interview that I conducted with the Australian author Greg Egan in the science fiction magazine *Aurealis*, Egan spoke of his approach to ideas such as virtual reality and artificial intelligence, referring specifically to his short story "Dust" (1992) and subsequent novel *Permutation City* (1994):

I recall being very bored and dissatisfied with the way most cyberpunk writers were treating virtual reality and artificial intelligence in the '80s; a lot of people were churning out very lame *noir* plots that utterly squandered the philosophical implications of the technology. I wrote a story called "Dust," which was later

---

[4] The other volumes are *Count Zero* (1986) and *Mona Lisa Overdrive* (1988).

expanded into *Permutation City*, that pushed very hard in the opposite direction, trying to take as seriously as possible all the implications of what it would mean to be software. (Blackford 2008, 19–20)

This attitude is doubtless shared by many writers working in the aftermath of the cyberpunk movement. They have incorporated much of cyberpunk's iconography while exploring its implications from unique viewpoints of their own.

Egan is one prominent author who has maintained a "hard science fiction" tradition, and there was an international resurgence of hard SF through the 1990s and beyond. This is showcased in David G. Hartwell and Kathryn Cramer's retrospective anthology *The Hard SF Renaissance* (2002). Unfortunately, however, the precise boundaries of hard science fiction are disputed. In their introduction to *The Hard SF Renaissance*, Hartwell and Cramer suggest that the expression "has always signified SF that has something centrally to do with science" (2002, 13). This "something" is vague, but hard science fiction generally emphasizes logic, scientific accuracy, and plausible detail. It often includes realistic descriptions of the lives of working scientists, engineers, astronauts, and similar professionals—characters who are imbued with a spirit of empirical inquiry, as they attempt to solve problems within known scientific principles. The hard SF sub-genre is epitomized by such works as Hal Clement's *Mission of Gravity* (1954), Arthur C. Clarke's *Rendezvous with Rama*, and Gregory Benford's widely-admired story of communication across time, *Timescape* (1980).

The most explicitly "hard" forms of hard SF show a line-by-line concern for scientific plausibility and/or a central concern with scientific knowledge to solve a problem or resolve a crisis. These features, in particular, distinguish it from space opera, stories of social extrapolation, and the futuristic technothriller subgenre. Such features are especially important in stories of survival or rescue in extraterrestrial environments. For the author, part of the challenge is to portray the characters reaching useful solutions within established constraints. Any solutions obtained must be consistent with actual science, or at least with whatever alternative scientific principles are established as operating within the world of the narrative.

In the US, the renaissance of hard SF in the 1990s was strongly associated with the work of Benford, Greg Bear, and David Brin, and with Kim Stanley Robinson's Mars trilogy—*Red Mars* (1993), *Green Mars* (1994), and *Blue Mars* (1996)—but numerous other writers also took part, including newer talents such as Ted Chiang. The renaissance was also evident in other English-speaking countries. It can be seen, for example, in Egan's stories and novels, and in the work of a group of British writers who include Ken MacLeod and Stephen Baxter.

Since the 1980s, no single movement in science fiction has reshaped it or challenged its existing expectations in the way of the New Wave or the phenomenon of cyberpunk. However, older forms have survived or been revived, and there have been new experiments, sometimes further blurring science fiction's borders with modern fantasy (on one hand) and with the literary mainstream. During the 1980s and beyond, James Morrow produced darkly satirical narratives, combining SF and fantasy elements, while Connie Willis became widely acclaimed, primarily for work that gave new interest and vitality to the tradition of time-travel narratives. She began writing professionally in the late 1970s, but emerged as a major talent during the following decade. Her 1982 novella "Fire Watch" won both of the science fiction field's premier awards—a Hugo Award and a Nebula Award—and it remains one of the most admired of all time-travel stories. Willis has since won many awards; arguably, she is the most professionally honored author ever to work in genre SF.

From the 1980s onward, Gene Wolfe has revived far-future science fiction in several novel series that can be read as one oceanically enormous narrative. This commences with *The Shadow of the Torturer* (1980), which tells the story of Severian, a journeyman torturer who is ultimately made Autarch. He must find a way to renew the dying Sun, for which he travels to a higher universe, Yesod, and pleads with an entity closer to the Increate.

Another notable development has been the revival of space opera in the medium of printed prose. (Looking beyond the print medium, space opera's surge in popularity began with the lucrative example of the first Star Wars movies in the late 1970s and early 1980s.) During the 1980s and 1990s, space opera's literary revival, sometimes discussed as the New Space Opera, was led by such varied writers as C.J. Cherryh, Catherine Asaro, Dan Simmons, David Brin, Elizabeth Moon, Iain M. Banks, Lois McMaster Bujold, Orson Scott Card, and Vernor Vinge. Others, such as Alastair Reynolds and Ann Leckie, are more recent additions to this group. These and many others took space opera in dazzling new directions. Their stories are set against the wide backdrops of the space opera sub-genre, but with emphases that range from intellectual speculation and play to action-adventure to unashamed romance. Much of their work has a strongly cyberpunk or post-cyberpunk sensibility.

The healthy state of space opera continues in the second decade of the twenty-first century, with novels such as those in Leckie's Imperial Radch series (commencing with *Ancillary Justice*, 2013) winning major awards. Leckie's work, too, has a prominent post-cyberpunk element. Even as the decades pass, the spirit and iconography of cyberpunk have made their mark not only on the entire science fiction field but also beyond into the more general art and culture of industrialized nations.

Science Fiction in Other Media

As the pace of social and technological change accelerated during the twentieth century, narratives of technological innovation and futuristic prospects became more culturally visible. Science fiction expanded into new media such as radio, cinema, comics, television, and computer games. The success of SF movies and TV series since the 1960s and '70s added greatly to the genre's audience—while, perhaps, not doing justice to its richness and variety.

The history of science fiction comics begins with the space opera comic strips introduced into newspapers in the early decades of the twentieth century, such as the *Buck Rogers* strip first published in 1929 and the *Flash Gordon* strip first published in 1934. These led to reprints in comic-book form and then to original comic-book series, such as *Action Comics*, which introduced the figure of Superman in 1938. By the early 1950s, SF-related comics were popular, with an emphasis on interplanetary adventure and space opera, though a broad range of science fiction types and themes was covered, and there were frequent adaptations of prose science fiction by well-known authors of the time.

While the world of comic books is complex, the general tendency since the 1960s has been for American comics, in particular, to be dominated by superhero adventures. Two major companies, DC (Detective Comics) and Marvel Comics lead the English-language superhero market, though they do have competition from others. Between them, DC (which publishes comics featuring Superman, Batman, and Wonder Woman, among a much larger cast) and Marvel have created countless superheroes and supervillains. Their adventures have fed back into televized, cinematic, and prose SF, with numerous adaptations of superhero comic series. Some cinema adaptations, such as those involving Superman, Batman, and Wonder Woman, and Marvel Comics' popular heroes, such as the Avengers and the X-Men, have achieved high levels of commercial success, often setting box office records. These movies have inspired further spin-offs, including novelizations of film scripts and entirely original superhero adventures in prose form.

Meanwhile, someone better qualified than I could write a separate book on SF-related manga produced by Japanese creators, or on manga-influenced and SF-related comics produced elsewhere in the world.

Cinematic science fiction developed in the early decades of the twentieth century, sometimes showing a dystopian edge, as in *Metropolis* (dir. Fritz Lang, novelization of the screenplay 1926; cinematic release 1927). During the 1930s, there was something of a boom in American science fiction cinema. This included James Whale's 1931 version of *Frankenstein*, which simplified (not to say dumbed down) Mary Shelley's novel, and adaptations of works by H.G. Wells. There were also entirely new stories, as with *King Kong* (dir. Merian C. Cooper and Ernest B. Schoedsack, 1933). As everyone knows, *King Kong* is a lost-world adventure: it features a prehistoric landscape on a remote island, an isolated tribe that practices human sacrifice, and a gigantic gorilla that falls in love with beautiful Ann Darrow (memorably played by Fay Wray).

Some early SF movies were optimistic about science and technology, most notably *Things to Come*, released in 1936. This was a British celebration of technocratic Wellsianism. However, the predominant emphasis through the 1950s remained Frankensteinian, with an evident suspicion of science and technology—most obviously in the many American and Japanese portrayals of monsters created by the abuse of science, such as the giant ants of *Them!* (dir. Gordon Douglas, 1954), and the huge, city-destroying reptile of *Gojira* (dir. Ishirō Honda, 1954) and its many sequels and off-shoots.

Other popular SF movies in the 1950s were *Destination Moon* (dir. George Pal, 1950), *The Day the Earth Stood Still* (dir. Robert Wise, 1951), *The Thing from Another World* (dir. Christian Nyby, 1951), *When Worlds Collide* (dir. Rudolph Maté, 1951), *It Came From Outer Space* (dir. Jack Arnold, 1953), *The Creature from the Black Lagoon* (dir. Jack Arnold, 1954), *Forbidden Planet* (dir. Fred M. Wilcox, 1956), *Invasion of the Body Snatchers* (dir. Don Siegel, 1956), *The Blob* (dir. Irvin Yeaworth, 1958), and *The Fly* (dir. Kurt Neumann, 1958). As suggested by the titles, aliens, monsters, and space exploration provided the iconography of 1950s SF cinema. All of this was later parodied lovingly by Richard O'Brien's stage musical *The Rocky Horror Show* (first produced 1973) and its cinematic version, *The Rocky Horror Picture Show* (dir. Jim Sharman, 1975).

The 1960s and 1970s saw a large number of cautionary anti-technology movies, cinematic dystopias, and post-holocaust movies. Amongst all this, Stanley Kubrick's enigmatic *2001: A Space Odyssey* (1968), on which the director collaborated with Arthur C. Clarke, set a new benchmark. In more recent decades, adventure movies such as *Star Wars* (dir. George Lucas, 1977; now known as *Stars Wars: Episode IV – A New Hope*) have looked back to older forms of science fiction, such as superhero stories and Gernsback-era space opera stories such as those of "Doc" Smith, as well as drawing heavily on the conventions of heroic fantasy. *Star Wars* can also been seen as in the tradition of 1930s cinematic space opera, such as the 13-part film serial *Flash Gordon* (dir. Frederick Stephani, 1936).

During the late 1970s and the 1980s, a new style of science fiction cinema emerged. This began, perhaps, with *Alien* (dir. Ridley Scott, 1979), and it included that film's sequels, plus *The Terminator* (dir. James Cameron, 1984), David Cronenberg's re-make of *The Fly* (1986), *Predator* (dir. John McTiernan, 1987), and the sequels to each of these. This was a darker, grittier—and sometimes distinctly ickier—approach to science fiction. To some extent, it was rooted in longstanding traditions of technophobic horror, but it also had affinities with contemporaneous developments in prose SF— that is, with the cyberpunk movement of the time—and in superhero comics, where a darker style became prominent during the 1980s (associated, in particular, with the work of Frank Miller).

Blade Runner (dir. Ridley Scott, 1982) was based on Philip K. Dick's *Do Androids Dream of Electric Sheep?* Like Dick's novel, *Blade Runner* depicts a police hunt for renegade androids in a near-future Los Angeles. Appearing at about the same time as the first clearly cyberpunk short stories by Gibson and others, and anticipating *Neuromancer* by two years, *Blade Runner* is pure cyberpunk, imagined for the big screen. Its importance as a masterpiece of cyberpunk filmmaking is rivalled only by *The Matrix* (dir. The Wachowskis, 1999) and its sequels.

As I write in 2017, SF and SF-related movies now dominate at the box office. These often take the form of space opera, superhero narratives, monster movies, and stories of alien invasion, though current moviegoers with an interest in science fiction are treated to a variety of intelligent movies that fall outside these categories. Dating back over several decades, SF cinema has also included numerous movies dealing with artificial intelligence, often of the "machines will take over" variety. *2001: A Space Odyssey* is notable for many reasons, but one of them is its beautifully choreographed depiction of conflict between human astronauts and their spaceship's advanced computer, HAL 9000. Stanley Kubrick returned to the theme of artificial intelligence with the poignant *A.I. Artificial Intelligence* (2001), which was ultimately directed by Steven Spielberg after Kubrick's death. Recent variations on this theme include *Her* (dir. Spike Jonze, 2013), *Transcendence* (dir. Wally Pfister, 2014), *Chappie* (dir. Neill Blomkamp, 2015), and *Ex Machina* (dir. Alex Garland, 2015)—all exhibiting quite different attitudes and tones.

Science fiction concepts first appeared in radio serials from the 1930s: *Buck Rogers in the Twenty-Fifth Century* (1932–1947), a space opera serial aimed at a young audience, was probably the first genuine science fiction radio program. Science fiction radio drama is now rare, but it was relatively common from the

1930s to the 1950s. Broadcasts usually took the form of juvenile serials, including the *Superman* series (1940–1952), in which the superhero from Krypton foiled the schemes of various criminals. However, the most famous radio drama of all was a 1938 broadcast by Orson Welles: an adaptation of H.G. Wells's 1897 *The War of the Worlds*. This produced widespread panic when many listeners mistook it for a genuine news report about an invasion from the planet Mars.

Science fiction television dates back to the late 1940s. The first series to be broadcast in the US was the post-war *Captain Video* (1949–1953, 1955–1956), which featured a futuristic hero who battled various alien threats. It was followed by many space adventure and superhero programs, mainly aimed at young audiences. However, *The Twilight Zone* (1959–1964) and *The Outer Limits* (1963–1965) were anthology-style programs with more appeal to adults. The former used a mix of SF and fantasy, while the latter was focused entirely on science fiction, often reflecting something of the New Wave sensibility of the time. Some of its most celebrated scripts, including "Demon with a Glass Hand" (1964), were written by Harlan Ellison.

The 1960s also saw the production of many science fiction or fantasy comedy shows, such as *My Favorite Martian* (1963–1966). Drama series of the 1960s included *Voyage to the Bottom of the Sea* (1964–1968), *Lost in Space* (1965–1968), and *The Time Tunnel* (1966–1967), all aimed at children and teenagers, and *The Invaders* (1967–1968), which sought a more mature audience. *Star Trek* (1966–1969) maintained a strand of optimism in the 1960s, at a time when science fiction in printed form was rather the opposite of optimistic. The episode widely regarded as the best in the original series of *Star Trek* was again one scripted by Harlan Ellison: "The City on the Edge of Forever" (1967). Though it did not attract strong ratings, *Star Trek* gained an enthusiastic fan base and eventually had many successors and imitators, as well as becoming the basis for a long succession of movies, commencing with *Star Trek: The Motion Picture* (dir. Robert Wise, 1979), and for numerous spin-off novels and comic books.

The show's cult following, together with the success of the Star Wars movies, led to the creation of many science fiction TV series in the 1970s and beyond, with fluctuations in priorities and fashions. From the 1960s, when SF programs on television were rare—and it was reasonably possible to keep up with all of them—SF has gained in popularity until it seems to be everywhere on twenty-first century television. Likewise, there has been an explosion of SF narratives and settings in computer games of all kinds: a booming entertainment medium that has now gained enormous participation, especially among younger people, and become a wealthy industry in its own right.

Concluding Remarks

Cultural ubiquity and bewildering variety characterize twenty-first century science fiction. The contemporary genre includes a teeming diversity of movements, manifestos, styles, and sub-genres, not all of which I have been able to mention in this relatively short historical survey. Many forms thrive; many boundaries are blurred. Space opera in print form shows influences from hard SF and cyberpunk. It often displays a higher degree of realism, and a greater distance from fantasy, than its cinematic equivalents. The mega-text of SF icons and tropes has metamorphosed over time into something darker, harder, and weirder.

There is now an intricate cross-fertilization of cinema, comics, television, media-related prose science fiction, and computer games. Along with the genre of modern fantasy, science fiction has become a conspicuous, even dominant, presence in international popular culture. But despite this pop culture visibility, or perhaps because of it, the kinds of science fiction that are best known to the public do not accurately reflect the state of the art in prose SF. Cinematic space opera typically resembles the stories that first appeared in the 1920s and 1930s. The relatively peripheral superhero variety of science fiction has obtained great prominence in the cinema, while more thoughtful movies with SF characteristics are scarcely marketed or discussed as science fiction at all. Think, for example, of *The Eternal Sunshine of the Spotless Mind* (dir. Michel Gondry, 2004).[5]

There has been no true merger of science fiction and the literary mainstream, often forcing writers to choose a career in either one or the other. Many literary authors have published stories and novels that are formally science fiction, or close to it: Margaret Atwood is an especially high-profile example. But SF has also become a major component in a popular culture whose values are often remote from those of more "literary" authors who emphasize psychological realism.

One exciting development in recent decades has been the talent explosion in countries where English is not the main language spoken and written. In Chapter 1, I acknowledged that I do not feel qualified to discuss this in any adequate depth. However, I also mentioned that the Chinese writer Cixin Liu (Liu Cixin) has become one of the genre's leading practitioners on a world scale. His trilogy collectively entitled Remembrance of Earth's Past stands with any other major contributions to the SF canon. The novels concerned are *The*

---

[5] Though, to be fair, this seems to be regarded within the film industry as a "sci-fi comedy."

Three Body Problem (2006; English translation 2014), *The Dark Forest* (2008; English translation 2015), and *Death's End* (2010; English translation 2016). Liu's accounts of grand events during a war of two star systems are breathtaking, and he explores the ethics of desperate choices. The enigmatic scientist Luo Ji, the main character of *The Dark Forest*, can easily be seen as a successor to Asimov's Hari Seldon, who is referenced in the novel. Like Asimov's fiction, Liu's trilogy emphasizes the issue of who has the greatest depth of insight into complex, potentially catastrophic, situations.

As the twenty-first century rolls out, offering promises and surprises, scientific understanding and technological innovation continue to transform our world and even the ways in which we understand ourselves. As long as that is so, the implications of technoscience will fascinate writers from many artistic and other traditions. Many will choose SF as a means of self-expression, and many will employ its tropes and icons to engage with persistent questions that trouble the human mind.

References

Aldiss, B. W. (1973). *Billion year spree: The history of science fiction*. London: Weiden-feld & Nicolson.

Aldiss, B. W., & Wingrove, D. (1986). *Trillion year spree: The history of science fiction*. London: Gollancz.

Asimov, I. (1981). *Asimov on science fiction*. Garden City, NY: Doubleday.

Blackford, R. (2008). An interview with Greg Egan. *Aurealis: Australian science fiction and fantasy, 42*, 16–23.

Broderick, D. (1995). *Reading by starlight: Postmodern science fiction*. New York: Routledge.

Clareson, T. D. (1990). *Understanding contemporary American science fiction: The formative period (1926–1970)*. Columbia, SC: University of South Carolina Press.

Gunn, J. (2006). *Inside science fiction* (2nd ed.). Lanham, MD: Scarecrow.

Hartwell, D. G., & Cramer, K. (Eds.). (2002). *The hard SF renaissance*. New York: Tor.

Harris-Fain, D. (2005). *Understanding contemporary American science fiction: The age of maturity, 1970–2000*. Columbia, SC: University of South Carolina Press.

Hillegas, M. R. (1967). *The future as nightmare: H.G. Wells and the anti-utopians*. New York: Oxford University Press.

Suvin, D. (1979). *Metamorphoses of science fiction: On the poetics and history of a literary genre*. New Haven, CT: Yale University Press.

3

Morality, Science Fiction, and Enabling Form

Philosophy and Philosophical Questions

Etymologically speaking, philosophy is (translated from Greek) the love of wisdom. In practice, philosophers attempt the intellectually rigorous study of deep and perennial questions that defy human investigation by such means as science and historical scholarship. Although the findings of scientists and historians might be relevant to people inquiring into philosophical questions, something more usually seems necessary.

The questions of philosophy concern, for example, whether God—or perhaps a pantheon of gods—exists. Is there, or might there conceivably be, a transcendent world lying beyond reach of our senses? Conversely, can we demonstrate that anything at all exists outside our minds? Philosophers also question whether we are free, in some sense, to act as we wish, or whether our sense of having free will is an illusion. But what sort of evidence would be relevant, one way or another, to a question like that? Do we even know what we're talking about when we use language like "free will"? Philosophy quickly "goes meta"—it becomes self-reflexive—once philosophers start talking to each other about what philosophical questions *even mean*, how they could ever be answered, even in principle, and so on. Philosophers sometimes wonder whether the answers to some questions are closed to us, perhaps because of our intellectual or sensory limitations.

Philosophy has changed over time. Large questions that fell within philosophy in ancient times, such as questions about the structure and composition of the universe, can now be studied scientifically. They fall outside of

© Springer International Publishing AG 2017
R. Blackford, *Science Fiction and the Moral Imagination*, Science and Fiction,
DOI 10.1007/978-3-319-61685-8_3

philosophy as it is understood today. But philosophers reflect on new issues when they emerge as social conundrums, such as issues to do with ethics in health care and research and issues to do with racial identity.

Crucially for the purposes of this book, philosophers examine questions relating to morality and ethics, including whether or not these are precisely the same concept (something that I'll discuss briefly in the next section). Among these questions, philosophers ask what is a good sort of life for human beings to live. They try to solve difficult moral problems, such as when, if ever, it's okay to sacrifice individuals for the sake of the common good. At another level—as always—philosophers wonder what criteria we could use to answer such elusive questions, and, indeed, whether the questions have answers at all.

Philosophical questions may seem baffling and speculative. Clearly, they are difficult to answer to everyone's satisfaction; worse, they are difficult to formulate without controversy. It's not clear whether philosophers have made progress with them over the last, say, two and a half millennia.[1] We might suspect that philosophical questions are misconceived, or meaningless, or futile, but most people do sometimes wonder about them, and it would be discomfiting to conclude that it's all just nonsense or that there are no correct answers to be found. Philosophers who draw that pessimistic conclusion might ultimately be correct, but they won't convince anyone without compelling arguments.

The world's religions offer answers to many of these deep, persistent questions, but *those* answers usually have to be taken on faith. Moreover, the answers given by different religions are, as often as not, in mutual contradiction, which might lower our trust in any of them. At any rate, it would be good to know how far human reason can take us in addressing philosophical questions, unaided by any non-rational "ways of knowing" such as claims to have received messages from angels or God.

A Very Brief Introduction to Moral Philosophy

For my purposes, moral philosophy and the branch of philosophy called "ethics" are the same thing, which is not to say that morality and ethics in the everyday senses are the same. Unfortunately, words such as "morality," "moral," "ethics," "ethical," and so on, are used in a bewildering variety of ways. I doubt that any of us are totally consistent in our everyday use of these

---

[1] For more, see the various contributions to Blackford and Broderick (eds.) 2017.

words, but morality usually has something to do with a society's *mores*: its most basic and strongly enforced standards of conduct (see "Mores" and "Norms" in Calhoun 2002). Talk of "ethics" often suggests something more closely based on reflection and reason—as opposed to whatever are the local social or religious standards—and it can refer specifically to the ancient philosophical tradition of asking how we ought to live our lives. In common parlance, it can also mean a code, presumably worked out in some rational way, by which a profession should be practiced, as with "legal ethics" or "medical ethics."

Viewed from an anthropological perspective, morality is a social phenomenon that can be studied in a detached, objective way. Anthropologists can investigate what standards of conduct are applied by the diverse range of human societies, including societies very different from our own. Similarly, though usually through rather different methods, historians can examine what standards of conduct were adopted at various times in human history. Sociologists might wish to study what standards are accepted in their own societies, perhaps identifying different standards among different demographic groups. All of this can reveal a great deal about the standards used by various societies and sub-groups. For example, human social groups do not operate only with codes of obligatory and forbidden conduct. They usually have some more flexible standards that include what are often called moral virtues. These are socially approved dispositions of character relating to the sorts of choices that individuals make in situations where choice is called for. For example, most societies regard courage, generosity, and fairness as moral virtues. The opposite qualities are vices of character (or moral vices), such as cowardice, meanness, and unfairness.

The local standards are also likely to have some flexibility in the sense that many actions are neither obligatory nor forbidden, which leaves a degree of personal choice. Even within this zone, however, some actions might be considered praiseworthy, though not strictly obligatory—philosophers call these supererogatory acts. Other actions might be frowned upon or considered shameful, although they are not strictly forbidden. A society's standards of conduct can be enforced through a variety of mechanisms. These can include promises and threats of divine judgment in a spiritual afterlife (or of consequences for future incarnations on Earth). Important standards of conduct can take the form of laws with formal punishments attached to their breach. However, the local standards are also enforced via less formal mechanisms. These can range from disapproving glances to public shaming and outright ostracism.

All of this is fascinating, but it does not settle the questions that most ordinary people and most philosophers worry about. Irrespective of what the local standards might say, how should I *really* live my life and act in various puzzling situations? Furthermore, given the dazzling variety of moral standards—standards of acceptable or unacceptable conduct, and of what counts as good or bad character—which are actually the *true* or *best* ones? Should we live our lives by a system of standards worked out through reason and reflection—inevitably something more abstract and theoretical than the standards of the local society—and if so what might it be? We might also be led to ask a more skeptical question: are there any rationally justifiable moral standards at all?

Patricia Churchland, the renowned neurophilosopher, makes the distinction between an anthropological and a philosophical approach to morality in this way:

> Mainstream moral philosophers tend to regard the description of a culture's social rules as of mainly anthropological interest, and not at the heart of morality in its profound, normative sense—what rules *ought* to be followed. (2011, 184)

Philosophers are reluctant to study morality in a merely descriptive way, much as they wish to be informed by the findings of anthropologists, historians, sociologists, and others. Moral philosophers spend much of their time studying the anthropology, history, and sociology of morality. But in the end, they want to sort out the content of the true or best moral system, and/or to establish what deeper theory might guide how we live our lives irrespective of the local norms.

Moral philosophers also ask themselves secondary "metaethical" questions: such as what is meant by moral language and whether we should believe that *any* moral system has cross-cultural validity. Some philosophers come to pessimistic or deeply skeptical conclusions about morality. Some deny that there is any true or best moral system or ethical theory. To be sure, skeptical doubts often occur to ordinary people who are not philosophers. There can be—or seem to be—something slippery or fishy about morality, as if it makes demands on us for which it has no genuine authority.[2]

[2] For an introduction to this topic, see *The Mystery of Moral Authority* (Blackford 2016), 1–9. This book can be used as a general introduction to moral philosophy, though from a somewhat skeptical perspective.

Theories of Ethics: Deontology, Consequentialism, Virtue Ethics

For many people, questions of morality are closely tied to religion. For these people, the true moral system might be whatever standards of conduct are set out in relevant holy books and authoritative commentaries on them by theologians, church leaders, and similar thinkers. Religious moral systems often include codes of absolute rules, especially absolute prohibitions. In that case, they are examples of the *deontological* approach to morality and ethics: that is, an approach where morally right conduct amounts to obeying a code that specifies which actions are obligatory and which are forbidden.

In the late eighteenth century, the German philosopher Immanuel Kant produced the most influential intellectual defense ever attempted of a deontological theory of ethics. He tried to find rules that could be deduced purely from an exercise of practical reasoning. The general idea was to examine which "maxims" could be universally acceptable (Kant 1991 [1785]). That is, Kant asked his readers whether their rationale for behaving in a certain way was one that could be adopted by everybody in relevantly similar situations, or whether this would involve some kind of contradiction. According to Kant's approach, the moral law consists of whichever maxims survive that test.

For example, Kant thought he could justify broad and absolute rules against lying, stealing, and committing suicide, while also defending some vaguer, or "imperfect," obligations (obligations with a degree of choice in how we go about meeting them). On Kant's approach, we have obligations to develop and improve our own talents and to expend some effort and resources to help people in need. These obligations are genuinely binding on us, but there is an element of personal choice in *how* we might meet them.

Two hundred years later, Kant's arguments have attracted many defenses and many criticisms. It is fair to say that some of his arguments now seem more plausible than others. For example, many philosophers find his argument against the ethical acceptability of suicide rather tenuous. For myself, I doubt that Kant's approach, or anything like it, can be successful overall as a theory of morality.[3] Still, Kant had at least a limited point: if we are going to have rules as to how we must and must not act, not just *any* rules are possible. It would not, for example, be possible to have a rule that encourages or permits selfish lies, as the rule would immediately lose its point. In a society with such a rule, no one would ever be believed, so telling lies would be pointless. It is not,

---

[3] For the full story on why I think this, see Blackford 2016, 24–40.

however, so obvious that we can't have a rule permitting lies in certain specific circumstances, such as to prevent great harm to others or to fend off intrusive and impertinent questions about our private lives. This might still be consistent with a regime in which people are usually truthful and can generally be trusted to provide others with accurate information.

Deontological approaches can be contrasted with consequentialist approaches to ethics. Here, the aim is not to follow any enumerable code of rules but to aim at certain consequences. The best-known consequentialist approach—that is, the best-known variety of *consequentialism*—is *utilitarianism*. Alas, this itself comes in a number of different forms and flavors. However, the general idea is to aim at the greatest overall happiness: perhaps of all people in your society, perhaps of all people in the world, or perhaps even of all living things in the universe. On a utilitarian approach, a society's laws, *mores*, and other standards of conduct are open to ethical criticism based on whether or not conforming to them produces (or perhaps even hinders) overall happiness. Not surprisingly, the great utilitarian philosophers of the past, such as Jeremy Bentham and John Stuart Mill, were active in efforts at law reform. The same applies to contemporary utilitarian philosophers, such as Peter Singer.

A tricky question for utilitarianism is how we are to understand happiness. Philosophers have identified problems if happiness is interpreted as meaning simply feelings of pleasure. Utilitarianism is often given a make-over and reinterpreted as about having our preferences satisfied, even if these are not necessarily preferences for pleasant feelings. I might, of course, have a desire for feelings of pleasure within my set of preferences. I might certainly have a strong preference *not* to experience agonizing pain! But there might be many other things, as well, in my set of preferences. I might, for example, want to have an accurate understanding of the world around me, even if what I learn is disconcerting. I might want to achieve certain kinds of success or to have a life that is flourishing by some standard (of course, such standards are themselves controversial). I might have a preference that certain things happen after I die, even though I won't be around to take pleasure in them.

All of this might still seem vague, and there are difficulties in understanding how there can be specifiable quantities of preference satisfaction. How are we supposed to measure and compare the strengths of different people's preferences? Can we assess the (likely) consequences of one action against another, and so conclude with any precision which action will (probably) bring about the greatest overall preference satisfaction? We don't even seem to have a natural unit of measurement for all this.

But it is also difficult to see the point of morality and its observable manifestations—such as the local social *mores* and practices for enforcing them—if it is not to help get us things that most people want, such as prosperous societies, happiness for more people, and avoidance of pain and suffering. We see this when governments engage in social planning and then enact laws, or spend public money, accordingly. Government bureaucracies cannot accurately measure anything as vague as happiness or preference satisfaction, but they use what might seem like reasonable proxies. They might, for example, try to predict the overall economic effects of certain policy options. In an area such as public health, the government might try to work out which policies would save the most lives, or perhaps add the greatest overall number of years to people's lives. The latter might be weighted by some measure of the quality of life in the added years.

Consequentialists are not always utilitarians. Some might want to maximize what they regard as goods other than happiness, preference satisfaction, or anything else that a utilitarian could plausibly have in mind by "utility." These non-utilitarian consequentialists might, for example, think it more important to create beauty than to maximize happiness. Some thinkers of a generally consequentialist inclination might even think that, at bottom, ethics is egoistic. For them, the idea is not to maximize overall happiness—even at the level of your local society—but to maximize *your own* happiness or to satisfy *your own* desires. Other thinkers might take an eroscentric approach: the idea is to maximize the happiness, or satisfy the desires, of the people whom you happen to love and care about. Paradoxically, these seemingly selfish approaches might not necessarily lead to behavior that is usually regarded as morally vicious. Why not? One reason is that an ethical egoist might reason that an effective way to achieve her own happiness is to be considerate, honest, and generous in dealings with others.

In recent decades, moral philosophers have given much attention to reviving ancient approaches to ethics based more on individual moral virtues and vices than on prohibitions and obligations, or on consequentialist reasoning. These approaches focus on dispositions of character that affect our choices in life. This philosophical project usually goes by the name of "virtue ethics." Virtue ethics might be especially relevant when reading novels, or watching movies, when we very often make moral assessments of the various characters that writers have created for us.

In *Ethics, Evil, and Fiction* (1997), the British philosopher Colin McGinn urges the philosophical study of fiction from a viewpoint grounded in moral philosophy, and especially in philosophical thought relating to moral character. Much of McGinn's book studies spiritual beauty as it is represented in works of fiction. On the other hand, McGinn considers the idea of evil beings in a particular sense: humans or other intelligent beings who derive pleasure from the pain of others and experience pain from others' pleasure. These personalities are represented in English literature by, for example, Iago from Shakespeare's great tragic play *Othello*.

McGinn suggests that philosophers have been too influenced by what he calls the commandment style of moral discourse, and not enough by the "parable style" of writing (1997, 172): by texts that engage our informal understanding of human psychology, ask us to apply it in the context of a story's events, and so express a moral lesson. If we follow McGinn's approach, moral philosophers have over-emphasized rules that can be learned by rote, and they have under-emphasized engagement with morally compelling narratives, whether embodied in novels, movies, narrative poems, or other forms and media. In this engagement, says McGinn, we "draw upon an enormous background of tacit knowledge about human life not clearly codifiable into theoretical principles" (1997, 174).

Whether or not McGinn is fair about the efforts of other moral philosophers, some of his ideas ring true. As he elaborates, fictional narratives provoke us to engage with moral issues and to draw moral conclusions, especially when we condemn some characters and admire others. We should add, I think, that we oftentimes find ourselves wrestling with ambiguity when we find we *cannot* condemn or admire unreservedly. In the case of science fiction narratives, furthermore, we are likely to find ourselves condemning, admiring—or finding we *cannot* unreservedly condemn or unreservedly admire—entire societies, cultures, or even species. At this stage, let us refocus on the science fiction genre.

Science Fiction as Enabling Form

Philosophical questions—and not only questions in moral philosophy—can arise in many literary and other narratives, and most certainly in science fiction narratives. Science fiction can dramatize how our commonsense intuitions are placed under pressure by the findings of science, or merely by the demands of working out a consistent scenario for something as strange as time travel. This is also a common move of contemporary philosophers, who delight in bizarre thought experiments, so there are strong potential synergies between philosophy and science fiction. It's not surprising that many philosophers are fond of science fiction and that many science fiction writers appear to be mentally steeped in philosophy, whether by formal training or by wide reading and reflection.

In Ted Chiang's much-admired novella "Story of Your Life" (1998) (which was the basis for the movie *Arrival* (dir. Denis Villeneuve, 2016)), aliens arrive on Earth and depart with no explanation as to why. As often in science fiction, one of the themes here is humanity's epistemic limitations: that is, the limits on our ability to comprehend the immense and strange universe opened up by science.

The aliens in "Story of Your Life" have a four-dimensional viewpoint on reality—with time as the fourth dimension—and the main effect of their actions is to alter one character's perception of time. In learning their language she comes to understand the universe as a reality in which future events already exist, or are already set, so she has memories of things that have not yet happened. This sounds potentially paradoxical: what if she acted to try to avoid what is already set as happening in the future? But accompanying her knowledge of the future is a sense of obligation to act as foreseen. In this story, science fiction tropes—and especially the idea of intelligent aliens very different from ourselves—are used as enabling devices. They enable us to engage with, and explore, ideas of fatalism, determinism, free will, and the future. The SF trope of time travel can be used for similar purposes, as in *The Time Traveler's Wife* by Audrey Niffenegger (2003) and its cinematic adaptation (dir. Robert Schwentke, 2009). In a different book from this one, themes such as fate and free will might assume a greater prominence, and this would necessitate a somewhat different selection of works to discuss.

Connie Willis uses time travel for a rather different purpose. Her characters often visit harrowing periods in human history such as the time of the London Blitz ("Fire Watch") or the Black Death pandemic (*Doomsday Book* (1992)). Here, the emphasis is on the commonality of human experience and vulnerability across history—a point that is emphasized in *Doomsday Book* by setting up parallel problems in the past and the future. In "Fire Watch," first published in 1982, the time-traveling protagonist from a future Oxford University frequently wonders what he is meant to learn from his assignment to defend Saint Paul's Cathedral during the Blitz, and, moreover, what he is supposed to *achieve* during the process. In the end, the point of the assignment does not lie in any achievement but in the connections with individual people that he develops, and hence in his deepened sense that the individuals who lived and suffered in the past should not be viewed as mere statistics.

More generally, the shared iconography, or mega-text, of science fiction can provide the enabling form to dramatize a wide range of philosophical themes. I have borrowed the term *enabling form* from Donald L. Lawler, although similar language is commonplace in literary criticism. Lawler explains as follows:

> By "enabling form" I mean simply a contrivance which extends a writer's capacity for treating a subject or for expression beyond the limits of either traditional discourse or his own private views. To some extent, all fiction is an enabling form. I merely wish to emphasize at this point a construction which facilitates potential frames of reference beyond the ordinary conventions of his craft. (1977, 72)

This is somewhat abstract and obscure, but the idea is illuminated by Lawler's discussion of Kurt Vonnegut's second novel, *The Sirens of Titan* (1959). The science fictional aspects of Vonnegut's novel provide the events and images to support a philosophy of universal meaninglessness—and to consider an appropriate human response. The godlike (or at least angel-like, though not angelic) Winston Niles Rumfoord has been extended through space and time as a result of a kind of space-time warpage called, rather comically, a chrono-synclastic infundibulum.

Rumfoord is omniscient about past and (up to a certain time) future events in our solar system. He manipulates human civilization to create a new religion, but we learn that the alien Tralfamadorians have manipulated human civilization, including Rumfoord himself, in order to communicate with one of their messengers, Salo, who is stranded on Titan (one of the moons of Saturn). Thanks to the Tralfamadorians' manipulations, Salo eventually receives the replacement part that he needs for his spaceship, enabling him to complete his mission of carrying a trivial message of greeting across the universe. In the upshot, human history in its entirety has a meaning that can be "read" by a Tralfamadorian but is irrelevant to any human purpose. This seriously deflates human pretensions. It seems that we live in an absurd and meaningless universe; certainly, the course of our history has not been arranged in any way for our benefit.

Rumfoord's manipulations are more beneficent than those of the Tralfamadorians, even though he is ultimately manipulated by them himself. He designs a religion that will come to terms with a universe that gives human life no purpose or meaning. Thus, he founds The Church of God the Utterly Indifferent, dedicated to the belief that the chance advantages and disadvantages with which people are born should be voluntarily erased. Rumfoord succeeds in establishing his utopia on Earth, and it is given some sympathy in *The Sirens of Titan*. However, he is shown to be an untrustworthy hierophant at best and his Church of God the Utterly Indifferent is never presented as an unequivocal blessing. Rumfoord and his Church are distanced from us, and the Church is satirized rather than offered as a social panacea.

In addition, Rumfoord can be callous, unfair, willing to deceive, and even willing to shed others' blood to obtain his objectives. He instigates a war between Earth and an ill-equipped army that he creates on Mars, which results in the massacre of his army and just under two hundred thousand people killed or missing—though it all conduces to his purpose of uniting Earth's discordant peoples (Rumfoord is evidently a utilitarian). The mind revealed in quotations from various of Rumfoord's works shows a certain compassion for humanity in the abstract, but a failure to respect individuals as ends in themselves rather than as means to achieve his goals.

In fact, the ethical center of *The Sirens of Titan* is not Rumfoord's church or its set of doctrines, much as these are intended to improve human welfare. It is the initiation of two other characters into fuller humanity. They are Malachi Constant, a successful businessman who is put through extraordinary sufferings by Rumfoord, and Rumfoord's wife Beatrice. Constant begins as an aggressive, self-satisfied man who has never really loved anybody, despite a wealth of sexual experiences. Beatrice Rumfoord begins as an equally self-satisfied married virgin whose fastidiousness is represented emblematically in a painting of herself as a little girl dressed all in spotless white. By the end, these two are forced to take refuge in one another's society, and they come to understand the meaning of love. Beatrice had never outgrown the mentality of the little girl in the painting. On Titan, she finally accepts the mutual dependence of human beings in a universe that doesn't give a damn about us.

The Sirens of Titan is manifestly a work of science fiction, albeit a darkly humorous one whose scientific content is not to be taken seriously. It also offers a great deal of literary delight, combining a general accessibility with a tightness of design that makes every aspect an integral part of the whole. Successive readings of Vonnegut's novel reveal more and more of its control and complexity, yet it presents a surface that is immediately fascinating.

As a more recent example of SF engaging with ethics, the 2016 movie *Passengers* (dir. Morten Tyldum) portrays a starship on a long flight to a colony world circling a distant star system. As the action begins, all crew and passengers are held in suspended animation. After colliding with debris in space, however, the ship is damaged and some of its systems begin to break down. One passenger, Jim Preston—a mechanical engineer played by Chris Pratt—is woken from suspended animation as the ship continues toward its destination. This leaves him alone for the remainder of the 90-year journey, with no way to communicate with anyone on the ship or off it. He has no access to the crew in their hibernation pods, though he is able to inspect the pods of his fellow passengers and he eventually figures out how to wake them up. (Unfortunately, there is no way of restoring anyone, including himself, to suspended animation.) Driven to desperation by loneliness and the prospect of a life without human contact, Preston attempts suicide, but is unable to go through with it.

He becomes increasingly obsessed with waking one other passenger: the beautiful Aurora Lane, a journalist played by Jennifer Lawrence. Eventually, after a year on his own, he wakes her, immediately recoiling from the enormity of what he has done.

Preston is never in any doubt that he has done a terrible thing, taking away Lane's life as she'd planned it and potentially subjecting her to a life-in-death similar to what he'd been enduring. However, they bond, fall in love, and establish a passionate romantic and sexual relationship—built, unfortunately, on the lie that she also woke up inexplicably as the result of some kind of equipment failure. When she discovers the truth, Lane is inevitably outraged and distraught, even attempting to kill Preston at one point, though this time *she* is unable to go through to the end with a fatal act.

Passengers is complex and uncomfortable as it explores an ethically difficult area. Lane accuses Preston of, in effect, murdering her, and her reactions suggest that she feels violated. Built as it is on deceit, their relationship treads the borderlands of rape. But at the same time, we can wonder how we would have responded in a situation as desperate and terrifying as Preston's. The movie also makes clear that, however we characterize, or feel about, his crime, Preston is an essentially decent man—courageous, compassionate, and willing to sacrifice himself—as well as highly competent and resourceful. This is one of those cases where we can neither wholeheartedly condemn nor wholeheartedly admire.

Unfortunately, the final scenes detract from what makes the rest of the movie so intelligent and strong. Partly this is from a sudden lurch into physical danger, action, and self-sacrifice. Preston is severely hurt, and almost killed, but he is eventually revived unscathed. During this sequence, Lane is herself morally tested: Preston persuades her, in the face of her own desperation, to allow him to die—which would leave her alone for nine decades—if needed to save the rest of the passengers and crew. In itself, this sustains the film's complexity and ambiguity, although it tends to let Preston off the ethical hook by emphasizing that Lane is also vulnerable to temptation. Perhaps, too, Preston's near-death and symbolic resurrection are intended to absolve him of previous wrongdoing.

In the upshot the two main characters fashion an unequivocal happy ending. This—and along with it the restoration of the characters' romantic relationship—seems intended to vindicate Preston's original decision, removing the large ethical question mark that hung over it. All this undermines the movie's impression of moral or ethical seriousness.

Science Fiction and Moral Philosophy: Ideas and Examples

Most narratives that find publication include at least some moral content—some engagement with moral or ethical issues—but it is usually rather sparse. As James Gunn states, fictional narratives may do no more than reinforce the basic assumptions of the culture within which they are written, "such as that good will prevail or that good will prevail only if men and women of intelligence and character work at it hard enough" (Gunn 2006, 101).

In some cases, the point might be a less optimistic one about the need for decent people to do whatever they can under difficult circumstances, whether they ultimately prevail or not. This is at least part of the impact of *Cloud Atlas*, by David Mitchell (2004), and the movie based on it, directed by The Wachowskis and Tom Tykwer (2012). In both versions, *Cloud Atlas* offers a set of nested stories, located at different times in history. Its SF element is simply that the stories extend into the future, where the same imperative still applies. Philip K. Dick's *The Man in the High Castle* shows a similar need for simple decency in a world where the Axis powers prevailed in World War II.

In *New York 2140* (2017), Kim Stanley Robinson portrays a future world ruined by the effects of anthropogenic global warming. At one level, this is an explicitly political and didactic novel. It is not only a warning about greenhouse gas emissions and the plausible consequences of inaction, but also a critique of capitalism and especially financial corporations. At another level, however, it shows how human societies struggle to survive in a catastrophic global situation, and, again, how some characters retain their decency despite the extreme circumstances and their own flaws. Part of the enjoyment comes from Robinson's leisurely weaving together of narrative strands, allowing an unlikely group of individuals to build alliances and friendships, and in some cases to show unexpected virtues of character. Among them, Franklin Garr, an impatient, self-absorbed, thirty-something finance trader, gradually reveals more depth than we or he might have expected.

Gunn adds that some fiction "attempts to say something more" than a simple message that goodness will prevail, or that it will prevail if we work hard enough at it. It may tell us "something more—about the nature or goal of life, the nature or difficulties of society, or the nature or problems of people: that is, something about the human condition" (2006, 101). In principle, science fiction is well placed to tell us that "something more." It assumes the mutability of current knowledge, technologies, and social forms. It often portrays technological and other novelties that enable its characters to act within constraints—physical, biological, technological, or cultural—that differ from those applying to human beings in contemporary and historical societies. As I described in Chapter 1, this means that science fiction might be better understood as a narrative *mode* rather than as a literary *genre*. It is defined, in large part, by its particular relaxation of constraints on what events can be depicted and what actions its characters can perform. All of this can be deployed for a wide range of effects.

In some cases, SF tropes enable the exploration of specific aspects of human behavior or the human condition. Science fiction enables authors to speculate about the future or to comment on the customs and values of existing societies. Thus, the genre has lent itself to the purposes of utopian or visionary speculation, and to various kinds of satire. It can, for example, extrapolate the worst tendencies of a real society, or it can view a real society from the perspective of an imaginary one with different assumptions.

Even future utopias are sometimes presented as imperfect, which, in itself, conveys a message about worrying tendencies in human societies. In *The Dispossessed*, Ursula K. Le Guin presents the civilization created on Anarres, a moon that orbits the planet Urras within the star system of Tau Ceti. Here, a group of anarchists has created a civilization along lines that avoid anything like a traditional state apparatus, a code of criminal law, or a monetary system. There is a decentralized planning system, but it relies on a widely internalized ethic of mutual aid and voluntary social contributions. All of this seems to invite our approval as far as it goes.

As we soon discover, however, the Anarresti ethic has degenerated. While most people retain the old ideals, some manipulate the system to selfish ends, and the individual commitment of social contributions has become, in large part, a matter of conformity to public opinion. Even individuals who notice what is going on, such as the brilliant scientist Shevek (and, even before him, his friend Bedap) find themselves constantly required to defer, dissimulate, and compromise. *The Dispossessed* shows how high-minded ideas can fall short of reality—inevitably, it seems, they are undermined by fanatics and self-serving manipulators, and by unforeseen contingencies—but we are also shown that all is not lost. The book's eventual emphasis is on the Anarresti society's resources for permanent revolution and self-renewal. Despite all, the final mood is one of optimism.

Science fiction narratives often include characters whose abilities differ from those of historical humans, sometimes because they have greater inherent capacities, and often because of their access to innovative and powerful technology. At one extreme, this encourages the creation of superheroes and supervillains—good and evil characters with superhuman abilities—and science fiction often relies for its effect on the spectacle that ensues when they are drawn into violent conflict. Similar spectacles come about when SF narratives include wars and battles between rival military forces equipped with super-weapons. As discussed in Chapter 1, such forms of spectacular combat—often found in planetary romance and space opera—create an affinity with heroic fantasy, where magical locales and characters are deployed for similar spectacular entertainment.

Insofar as a serious purpose emerges here, it is in dramatizing issues about the use of great power, a theme that I'll explore further in Chapter 5. Powerful individuals often need to make choices about how to *use* their power: for self-aggrandisement or for the benefit of others? This is highlighted throughout the Star Wars movies, for example, when potentially powerful individuals such as Anakin Skywalker and Luke Skywalker are faced with choices to use the power of "the Force" for good or evil. The recent *Rogue One: A Star Wars Story* (dir. Gareth Edwards, 2016) retains something of this emphasis, although it differs from earlier movies in the franchise by offering a different picture of heroism.

Rogue One is peripheral to the main Star Wars series: its events occur immediately before those of the original *Star Wars* (now known as *Star Wars: Episode IV—A New Hope*), which really needed no further explanation. There is much fantastical action involving space battles and super-weapons, and like all Star Wars movies *Rogue One* is unashamed space opera. It has a cast of characters who are distinctly larger than life, yet they are considerably less so than those in the other Star Wars movies. Their important choices are not about how to use the Force, but they are nonetheless enormously consequential. Among them are Galen Erso's decision to build a planet-destroying Death Star (while also incorporating a weakness so it can be attacked successfully, as seen in *Star Wars*), Jyn Erso's decision to join and fight for the Rebel Alliance against the oppressive and morally corrupt Empire (even though involvement with the rebellion has previously given her nothing but grief), and Cassian Andor's decision to disobey an order to assassinate Galen Erso.

Such choices are closer to ordinary human experience than the grander ones that confront, for example, Anakin Skywalker, known later in his life as Darth

Vader. But this is surely part of *Rogue One*'s point. It plays off the grand tapestry of the other movies in the franchise, demonstrating the more ordinary heroism and sacrifice of relatively ordinary people. It suggests what we really ought to have known, but Hollywood does not always acknowledge: heroism and hard choices exist at many levels of a glorious enterprise, and the outcome is often great sacrifice for those who commit themselves to a cause.

Human Nature in *Lord of the Flies* and *Tunnel in the Sky*

James Gunn's list of lessons that fiction might offer us, his "something more" than banal presentations of good people prevailing over not-so-good ones, includes lessons about the nature and problems of people—one component, as he states, of the human condition. Fictional tropes, not least those of science fiction, can enable our engagement with this. Consider William Golding's *Lord of the Flies* (1954). It is at most borderline SF, reaching that borderline in employing as its background a future nuclear war. But the minimal SF content is important to its theme, which involves a bleak view of human nature.

Golding's novel depicts the efforts and misadventures of a group of evacuated British schoolboys when they are stranded on an isolated island and attempt to build a workable society of their own. The island is as hospitable as anybody might wish, but the boys' efforts at cooperation soon fail. A few of the main characters make intelligent and well-meaning efforts, but the majority find it easier to let go of any veneer of civilization. They fall into superstition and barbarism. Many of the boys' problems stem from their childish fears and short attention spans. But with large-scale warfare taking place—including in aeroplanes high over the island—adult humans in the outside world are not doing much better than the children. The war among nations suggests that adults, too, fail at managing conflict. Though they might live in organized and viable societies, those same societies are aiming at each other's destruction. The implications for humanity's nature are disconcerting, not because we are inherently malevolent (some of the characters are thoughtful and decent, after all), but at least because cooperative society is a struggle for us to maintain.

Robert A. Heinlein's *Tunnel in the Sky* (1955) is a more straightforwardly science fictional narrative that responds to the sort of pessimistic scenario offered by Golding. Heinlein shows what happens when a group of well-trained, though inexperienced, teenagers and young adults are stranded away from civilization on a distant planet. Unlike the cultism and savagery that arise in *Lord of the Flies*, the result is a faltering, difficult, but eventually successful effort to build a productive society. Some of Heinlein's characters are brutal, selfish, lazy or merely foolish, but despite some false starts, the main character, Rod Walker, and his peers make efforts to establish a working constitutional government and an effective allocation of labor. Rod himself grows from a naive teenager with a propensity to jump to conclusions, based on ostensible "logic," and ends up as an effective leader of the settlement that they establish. Notably, the young women in the story tend to be at least as competent as their male counterparts—and less prone to selfishness and silly mistakes.

It is also noteworthy that Rod never becomes romantically involved with any of the young women, even though he becomes close friends with some and even though some of Heinlein's characters do pair off. He is also not especially physically heroic, and is even beaten badly in physical combat with one of the book's nastier characters. Though physical competence is portrayed positively in *Tunnel in the Sky*, it is not a book in which the ability to win physical encounters is assumed to be a great virtue. The virtuous do not necessarily prevail if it comes to outright fighting. But the majority of the characters recognize decency and well-meaning effort, and these are rewarded with loyalty from others. One character who initially undermines Rod politically—and even wrests the leadership from him for a time—is revealed as not villainous but as doing his best, and as not entirely foolish or ignorant.

Although difficulties are encountered and lives lost, Heinlein's young people muddle through the problems with whatever skills, training, and sense of mutual commitment they can muster. Again, these are rewarded because the characters show enough sense to recognize and value them in each other. There is an element of frontier romanticism here, but it is tempered by doses of realism. When the stranded group is finally rescued, the adults assume that the young people must have fallen into barbarism, even though nothing so simple, or anywhere near so extreme, ever happened. Overall, Rod and the others make a good fist of things.

James Blish: *A Case of Conscience*

James Blish's 1958 novel *A Case of Conscience* (based on a novella of the same name published in 1953) engages with questions in theology, religious morality, and metaethics, although it is first and foremost a psychological study of a devout man forced to make difficult choices in extraordinary circumstances. It leaves to the reader to decide whether its main character is deeply insightful ⋯ or obsessed and deluded.

As the novel begins, it portrays the dilemma of a four-man commission that has been assigned by the United Nations to investigate a planet, Lithia, 50 light-years from Earth. The commission's task is to recommend whether Lithia should be open to trade and other ongoing contact with Earth, or whether, at the extreme, it should be quarantined as too dangerous for human involvement. All four members of the group are scientists, although the main character, Ramon Ruiz-Sanchez, is a Jesuit priest as well as a highly expert biologist. Throughout the novel his foil is Paul Cleaver, a brilliant physicist who is nonetheless shown as coarse and venal. The other members of the commission are Michelis, a chemist, and Agronski, a geologist.

With half-hearted support from Agronski, who seems largely indifferent to Lithia's future and Earth's interactions with it, Cleaver proposes that the planet be reported as unfavorable for human contact. This would cut off trading relationships with the Lithians and other uncontrolled interactions. Cleaver sketches a plan to exploit Lithia's abundant supplies of lithium, available in the form of pegmatite. He argues that the planet is an ideal location for thermonuclear research and manufacture of thermonuclear weapons. Michelis brutally criticizes this plan. He favors opening the planet to wider human contact and particularly to extracting titanium, while Ruiz-Sanchez takes a diametrically opposite position based on theological worries. He recommends completely quarantining the planet from human contact including any exploitation of its natural resources.

Lithia is inhabited by intelligent, bipedal reptile-like creatures, which Cleaver refers to disdainfully as "Snakes." The Lithians are twelve feet tall and fearsome looking, but they have established what appears to be a utopian society. While Cleaver views them as a source of cheap labor, Ruiz-Sanchez has interacted with them more in a spirit of humility and developed respect and liking for them. Yet, as he outlines his reasoning to Cleaver, Michelis, and Agronksi, he draws an extraordinary conclusion: the planet is too good to be true. The Lithians have a society with a moral code that closely matches Christian—indeed, specifically Catholic—morality, yet no supernatural beliefs underpin it. The Lithians have no wars or similar conflicts, and they live harmoniously with nature. In many ways, then, their planet seems Edenic. From a theological viewpoint, it appears to be an unfallen world, but there is no concept of God.

Thus, Ruiz-Sanchez views Lithia as a moral trap set for humanity. It stands as an existence proof for something that the Church deems impossible. It seems to demonstrate that individuals, or an entire society or species, can be morally good by the Church's standards without any religious sensibility or beliefs. Viewed from outside, Lithian society stands for such heretical propositions (from the viewpoint of the Church) as that reason is always a sufficient guide to action, that the self-evident is real, that good works are ends in themselves, that right actions can exist without love, and that peace need not pass understanding. In brief, it seemingly shows that ethics can exist without evil alternatives, that morals can be sustained without feelings of conscience, and that moral goodness is possible without God. To Ruiz-Sanchez, all these ideas are proposals of Satan intended to undermine religious faith. And yet, Lithia shows them in action.

Ruiz-Sanchez also worries that ascribing such a trap to Satan's machinations suggests that the Adversary is creative in a sense denied by Catholic orthodoxy. Much of the story that follows depicts Ruiz-Sanchez's struggle to come to grips with this issue. He finds himself driven to thoughts about the abilities of Satan that the Church views as heretical. His new understanding of God and Satan is a more Manichean one than Catholic theology permits.

For my purposes, it is important to see how the issues raised by Lithia's mere existence provide a philosophical challenge. This extends beyond Lithia's contradiction to Catholic theology. In identifying a series of philosophical propositions as inherent in the nature and functioning of Lithia, Ruiz-Sanchez raises puzzling issues in metaethics—puzzles that are as difficult and troubling for secular philosophers as for theologians. While he views these propositions as theological heresies, some of them have been maintained strongly by moral thinkers in the tradition of Western philosophy. Several have a distinct Kantian ring, notably the idea that morality can be grounded solely in the exercise of reason.

Other philosophers have pointed out difficulties in basing morality on reason. Despite Kant's arguments and the contributions of many philosophers who have followed him, it has proved difficult, if not impossible, to ground moral claims in reason alone. Philosophers in the Western tradition have repeatedly shown an impulse to ground their ethical systems in something that seems immutable and deeper than our moral codes and intuitions. This might be God or it might be reason itself. But all attempts to carry out such a project are beset by problems.

This does not imply that morality is simply an illusion or that human societies could do without it. If we think of morality as a natural and social phenomenon—the observable phenomenon of various norms or standards of conduct found in current and historical societies—we might be able to understand why it is, in practice, an indispensable element of human life. We might explain it as a social technology that contributes to social cooperation. On this approach, then, it might be true that all human societies (and any comparable societies of intelligent aliens) need commonly accepted standards that restrain individual selfishness, opportunism, and malice, while encouraging actions of wider social value. For individuals inside a particular society, the local moral code will appear absolutely authoritative, but it will not appear so to someone (for example, an anthropologist) taking an external perspective. Nor is any moral code likely to gain the perfect allegiance of all concerned within the society on all occasions.

Among the Lithians, however, there appear to be no moral transgressions, even though transgressing would sometimes be in the interests of particular individuals. Ruiz-Sanchez points out to the other members of his UN-sanctioned group that the Lithians vary greatly among themselves and choose their own life courses. They are genuinely individuals, not hive or herd creatures, yet no Lithian ever commits an anti-social act and the Lithian language lacks a word for such acts. To a religious thinker, all this appears incredible in the absence of faith. So, Ruiz-Sanchez asks himself and the others, how do these entirely worldly and rational beings come to assume, and depend upon, fundamental moral claims that cannot be established by reason alone? As he puts it, it is possible to reason accurately *from* moral ideas such as equality before the law and sanctity of the individual, working out their implications and how they apply to particular cases. However, he insists, it is not possible to reason our way *to* such ideas.

In response, Cleaver is scornful. Michelis at first attributes the apparent perfection of Lithian society to its possession of enormously superior behavioral and social science. Later in the debate, however, he bolsters Ruiz-Sanchez's argument: from his own observations, he notes that much Lithian science appears to be based on ungrounded axioms that are not obvious, at least to humans, and may even contradict the assumptions made by human scientists. For Ruiz-Sanchez, the inescapable conclusion is that Lithia, and particularly the society of the Lithians, could not have arisen naturally. Lithia is being sustained by an external agency, which he identifies as the Ultimate Enemy—Satan himself.

As Ruiz-Sanchez gloomily expects, Cleaver's vision for Lithia prevails with the authorities back on Earth. Meanwhile, the UN group returns to Earth with a newly born Lithian, Egtverchi, who is then placed in the care of Liu Meid, a UN laboratory chief. Egtverchi grows rapidly to adulthood and soon rejects Earth society with disgust. He becomes a demagogue who calls on the planet's disaffected to revolt. Indeed, the Earth-based society portrayed in *A Case of Conscience* is dystopian, with the masses of humanity living in crowded underground shelters in fear of nuclear warfare. Although Egtverchi's revolt fails after 3 days of chaos, it stirs up a broader political movement that might yet produce some good. Meanwhile, Egtverchi himself flees to Lithia in an effort to escape the human authorities.

Ruiz-Sanchez fears what Egtverchi might accomplish as an apocalyptic demagogue on his own planet. In the end, however, Lithia is destroyed— seemingly by Cleaver's thermonuclear experiments. The entire planet explodes while being observed by a group assembled on the Moon for the purpose of watching its destruction. Among the group are Ruiz-Sanchez, Michelis, and Liu Meid, all chosen because of their various connections with Lithia. The great mathematical physicist Henri Petard has calculated an error in Cleaver's work that implies Cleaver will trigger a cataclysmic lithium-based chain reaction. Cleaver, of course, refuses to admit any such error. Accordingly, Petard calls his group together to observe an event that, as he explains, he profoundly hopes will not happen. They use a specially constructed telescope that enables them to observe events in real time even though they are many light-years distant.

Henri Petard is a pseudonym for the scientist's real name: Lucien le Comte des Bois-d'Averoign. The shorter version of his pseudonym, which he uses as pen-name, is H.O. Petard—presumably a joke meant (by Henri and by Blish) to suggest the Shakespearean phrase "hoist with [or by] his own petard," a petard being a small bomb. Although he is not blown up by his own bomb, Petard warns Cleaver that the latter risks something of the kind. The predicted chain reaction takes place, and Cleaver is therefore destroyed by his own experiments. In the moments before this, however, Ruiz-Sanchez exorcizes the whole planet, reciting a lengthy set of ritual demands for demons to depart Lithia. The planet's destruction follows almost immediately, producing a clear impression for the reader of cause and effect, as if the exorcism, rather than Cleaver's disastrous experiment, caused the explosion.

By placing Ruiz-Sanchez at the centre of the narrative, Blish makes the story a psychological one, focused on the priest's attempts to understand what he has observed on Lithia and to explain it in terms that make sense within his religious worldview. He eventually seems persuaded that the destruction of Lithia is both a natural *and* a supernatural phenomenon: that is, it is both the result of his prayers and the exorcism that he conducts from the Moon *and* the result of Cleaver's mistake that produces an explosion on a planetary scale.

Throughout the novel, Ruiz-Sanchez is obsessed by questions that might turn out not to have a single, straightforward answer. Just before he utters the words of the exorcism, he turns to a simple version of the sort of problem that so troubles him. He considers the common situation where critically ill children are saved by advanced medical science after receiving prayers for their recovery. Does the efficacy of the medical treatment mean that the prayers were unnecessary and played no role? In response, he falls back on a classical solution favored by the Church: a medical miracle, such as the healing action of a powerful antibiotic, is not unworthy of God's bounty. In other words, God answers prayers—but in his own way, which might involve natural processes. By implication, the destruction of Lithia by Cleaver's faulty experiment can also be the result of the exorcism. Natural and supernatural causes coexist.

This reasoning is not likely to persuade a skeptic, and Blish clearly leaves open a purely secular interpretation of the novel's events. Even Ruiz-Sanchez's argument—supported by reflections from Michelis—that Lithian society could not have come about naturally is hardly conclusive. Nothing that we are shown rules out a naturalistic explanation of how the Lithians and their culture evolved, and it is clear enough that the UN commission has studied this relatively superficially. Michelis's suggestion that the Lithians use advanced social science to ensure conformity with their moral standards is never ruled out, and for readers approaching the book years later Ruiz-Sanchez's whole approach is uncomfortably like the arguments from irreducible complexity that some religious apologists employ to "discover" intelligent intervention in the record of biological evolution. That is, Ruiz-Sanchez's approach resembles arguments that one or another biological system could not have evolved gradually from earlier, perhaps simpler, components. Most scientists have been unimpressed by this style of argument, and many allegedly irreducibly complex systems have been studied in detail and received plausible explanations.

Blish does not deal with any of this laboriously or explicitly. However, Ruiz-Sanchez never offers the other characters arguments that are sufficiently compelling for them to abandon their secular and naturalistic understandings of the world. In his own mind, Ruiz-Sanchez reconciles the destruction of Lithia by both natural and supernatural causes, but the other characters could surely view the supernatural explanation as redundant. It does no explanatory work that is not adequately performed by Henri Petard's calculations. There is no suggestion, even in Ruiz-Sanchez's mind, that scientific investigation of the explosion would discover any cause other than a chain reaction inadvertently set off by Cleaver.

The other characters on the Moon react variously to Ruiz-Sanchez's exorcism as it takes place. They are already awaiting the predicted destruction of the planet at any moment. A UN observer expresses irritation with Ruiz-Sanchez. Others stare in wonder at the screen, waiting to see whether the planet will, indeed, explode in the way Petard predicted. Among them, Liu

Meid shows fear, though it is not clear whether she is affected by the exorcism itself or solely by anticipation based on Petard's prediction. Petard appears knowing and solemn, but his attitude to the exorcism is inscrutable.

After the blinding destruction of Lithia, seen through the telescope, Petard leads the rest of the party away. They leave Ruiz-Sanchez alone with his faith in God and his grief at the horrific outcome of events. And there this complex, ambiguous, sensitive narrative concludes.

Mary Doria Russell and the Ways of God

In *The Sparrow* (1996) and its sequel *Children of God* (1998), Mary Doria Russell provides a more recent contribution to theological science fiction, although it has something of the ambiguity that I described in discussing *A Case of Conscience*. Russell portrays the first expeditions from Earth to the planet Rakhat, carried out by the Society of Jesus after the Arecibo Observatory in Puerto Rico receives radio broadcasts from the direction of Alpha Centauri. These take the form of beautiful songs, though they cannot be translated. Emilio Sandoz, a Jesuit Father and brilliant linguist from Puerto Rico, organizes the first expedition, drawing upon the talents of friends and colleagues and of other Jesuit priests.

Sandoz believes that his expedition is divinely inspired, but it proves disastrous for himself and his carefully chosen crew, while having cataclysmic ramifications on Rakhat itself, where the explorers from Earth come into contact with two species of intelligent aliens. These species are of similar intelligence and general appearance—but they have co-evolved with the fearsome Jana'ata as predators and the more gentle Runa as their prey. The human explorers first meet a village of the Runa, inadvertently introducing agriculture to a society that was based solely on tribal gathering. This transforms the economic base of Runa society, leading to a baby boom that is thereupon exploited by the voracious Jana'ata.

Though Father Sandoz had only innocent intentions, his actions in traveling to Rakhat and making contact with its intelligent species cause great suffering for himself and others around him. The havoc innocently caused by the arrival of humans on Rakhat is analogous to the effects of contact between Indigenous peoples and more technologically advanced Europeans during the centuries of colonial imperialism on Earth. The events also challenge our complacency by showing a much more modern, and seemingly enlightened, group wreaking similar havoc. Viewed in an even broader perspective, *The Sparrow* suggests that immense harms can result from the well-intentioned actions of thoughtful and morally virtuous people.

In the end, only Sandoz manages to return to his home planet, though he leaves behind the woman he has come to love, Sofia Mendes—unbeknownst to him, still alive among the Runa and expecting a child. Prior to his escape from Rakhat, Sandoz is imprisoned by the Jana'ata, who agonizingly mutilate his hands and subject him to months of savage, repeated anal rape. Back on Earth, where many years have passed in his absence (thanks to the time-dilating effects of space travel at relativistic velocities), Sandoz is subjected to an inquest by the Society of Jesus. The Jesuits originally assume that he allowed himself to be sodomized by the aliens. (One stream of the narrative, as Russell has structured the novel, depicts this humiliating inquest, while the other gradually shows us the events as they took place on Rakhat.) In a final mockery, as it seems, of Sandoz's naivety, the songs of the Jana'ata glorify exploits of rape.

Though Sandoz had good intentions, his actions in traveling to Rakhat and making contact with its intelligent species caused suffering for himself and others around him. As *The Sparrow* comes to an end, he cannot reconcile the course of events on Rakhat with the teachings of his faith, which describes a loving and providential God who, according to the New Testament Gospel of Matthew, is mindful of every tiny event such as the fall of each sparrow. At one level, then, *The Sparrow* is a book about the classical Problem of Evil: how can the existence of such a God, possessing infinite power to enact his intentions, be reconciled with the malice and suffering that we all observe in such profusion?

Children of God is markedly less bleak than *The Sparrow*, although Sandoz is again put through misery. Having lost his faith and left the priesthood, he wants nothing to do with any further expeditions to Rakhat. He even meets a woman, Gina, with whom he falls in love. She helps him regain something of his old strength and energy, and they plan to marry. Although he helps train a new crew to travel in a second Jesuit-organized expedition to Rakhat, he is adamant about not joining them. However, the Pope and other powerful officials of the Church continue to believe that his mission was divinely inspired and that he is destined to return to Rakhat. In a melodramatic turn of events, Sandoz finds himself beaten, kidnapped, and drugged by gangsters who are in league with the Church authorities. This takes place with the involvement or knowledge of the Pope himself.

And so, against his will, and to the destruction of the new life he has been able to establish on Earth, he does return to Rakhat. Here, he is confronted by the consequences of the first expedition and by ongoing power struggles on the planet.

Much of *Children of God* is a meditation on whether unjust deeds and episodes of great suffering can be redeemed by beneficial outcomes. This has implications for moral philosophy but also for theology and philosophy of religion. Marooned on Rakhat with her son Isaac, Mendes concludes that everyone involved with the first expedition—including the Jana'ata merchant, Supaari, who befriended them—made a mess of things. Though none of them were deliberately evil, they caused great harm. Likewise, Sandoz doubts whether any good results from the expedition and the conflicts it stirred up could ever justify the harms that were done. Even if the first expedition precipitated the liberation of the Runa, he insists, nothing about his interactions with the Jana'ata could provide fitting material for disquisitions on the sanctity of life or the benefits of freedom.

In the end, we might come to think that the disastrous expedition worked out for the best. Yet that question is never given a definitive answer. One strength of Russell's literary diptych of *The Sparrow* and *Children of God* is that it explores moral and theological anxieties, rather than presenting its readers with traditional dogma. If Russell manages to convey any positive message, she earns the right through her unstinting authorial acknowledgment of the harrowing evils that her diptych portrays.

Despite everything, and quite tentatively, Sandoz does eventually recover his lost faith. He is not much impressed by large events such as the liberation of the Runa, and with this the creation of a vibrant new Runa culture. But he's moved by the beautiful, unearthly music created by Isaac, who turns out to be an autistic savant as he comes of age. Isaac discovers his music within the blended DNA of Runa, Jana'ata, and humans. This naturally occurring, deeply hidden music seems miraculous, though as Sandoz himself realizes it is not conclusive proof of God's existence and action.

Mary Doria Russell has denied that Emilio Sandoz is based upon, or intended to allude to, James Blish's character Ramon Ruiz-Sanchez. In a statement on her website (Russell n.d.) she reveals that she was unaware of Blish's novel when she wrote *The Sparrow*, rendering any similarities coincidental. Her protagonist's surname actually came from a manufacturer's name on a medicine label. This is a good example of the perils of attempting to trace lines of influence within the science fiction genre (or in literature and art more generally). Similar ideas often suggest themselves to different authors with no line of direct, or even indirect, influence. It is interesting, nonetheless, to compare the approaches taken by Blish and Russell to similar artistic, philosophical, and theological problems. Both depict contact between alien civilizations and highly intelligent Catholic priests. Each protagonist undergoes life-changing experiences on an alien planet, and experiences extreme anxiety in the aftermath. Each finds his faith challenged, though in the case of Ruiz-Sanchez he is tempted to doubt not so much the loving providentiality of God as the limitations assigned by the Church to the power of Satan.

Both protagonists face much the same question about the ambiguity of God's actions. Can any event be miraculous if it is explicable in purely naturalistic terms? Ramon Ruiz-Sanchez is content that God can answer prayer through natural events such as the invention and timely use of antibiotics or the destruction of a planet through a failed experiment with thermonuclear power. Emilio Sandoz comes to a similar conclusion in the final pages of *Children of God*, when he regains his trust in God's presence and grace. Though the songs of the Jana'ata glorified rape, Isaac's songs are a far more pure form of music, discovered within nature itself and untainted by cultural ideas of any kind. Whether or not they are literally miraculous is another thing.

Dwelling on that question, Sandoz acknowledges to himself that there might be a natural explanation for Isaac's music. Perhaps a sufficiently extreme and obsessive musical genius could find music in almost any complex data, whatever its origin. Yet it seems sufficient for Sandoz that the music and other signs of hope point to God's mysterious involvement in the universe. This is not to say that *Children of God* (any more than *A Case of Conscience*) ends up being a work of Christian, or broadly theistic, propaganda. Although *Children of God* ends on a hopeful note, it remains in question whether all the humiliation, suffering, and destruction has ultimately been for good—and even if it has been, it is likewise left unclear whether good consequences redeem the evils and harms that led to them.

The Culture's Conundrum

During the 1980s and 1990s, space opera's revival in book form was encouraged by several ambitious novel series, among them the Culture series by Iain M. Banks, commencing with *Consider Phlebas* (1987). In fact, this is not so much a series as a setting for books and short stories that are largely independent of each other. The events described take place over a period of about 1500 years (depending on issues such as how we think about flashbacks and back story). These various novels and stories in the—well—series depict the problems of a vast space-faring society that is in many ways utopian. The Culture has a post-scarcity economy, has seemingly abolished all forms of discrimination and injustice, and allows its citizens virtually unlimited scope to engage in pleasure-seeking or devotion to their own personal projects. Most decisions of any significance to the society as a whole are handed over to powerful machine intelligences known as Minds, though some humans with exceptional talents do play an active part.

The Culture is frequently confronted by questions about how to engage with alien societies that do not share its values. Some of these societies are illiberal and harsh, and the Culture needs to decide whether to take an approach of non-interference or to involve itself for the greater good. Where practicable, it acts with discretion, attempting not to disrupt other societies. It tries to take minimal action even against barbaric, warlike, misogynist societies such as the Sarl in *Matter* (2008). In other cases, however, it is more forceful. In any event, as Damien Broderick and Paul Di Filippo summarize the recurring situation, "Many of the novels [in the Culture series] involve interstellar spies and manipulators known as Special Circumstances, and their harrowing moral quandaries" (2013, 58).

Often these manipulators are acting against the interests of brutal civilizations that rival the Culture, oppress their own people, and otherwise appear barbaric. But how far should the Culture go in opposing them, and by what moral authority does it act? There's the rub. Tricky questions of practical ethics can lead to even deeper, more troubling perplexities of a more metaethical nature. Thus, Patrick Thaddeus Jackson and James Heilman highlight the Culture's problem of moral uncertainty:

> The only worry that the Culture does have is that it does not have an objective viewpoint from which it can judge its actions. There is no entity that dispenses moral truths that can be known just as one would know about natural phenomena. At best, the Culture can believe that it acts rightly because it acts on principles that it thinks all should agree on. (2008, 248)

In *Use of Weapons* (1990), the third volume of the Culture series, the expression "use of weapons" refers to skill in bending everything to hand toward a single purpose, turning whatever resources are available into effective instruments. These might be, but are not necessarily, literal weapons. The novel's main character, Cheradenine Zakalwe, is expert in the use of weapons in this sense, but he is himself manipulated by the Culture to achieve its purposes. While those purposes might ultimately be for the universal good— and thus, perhaps, justifiable by a utilitarian calculus—the means employed are distasteful. They are especially damaging to individuals who are turned into the Culture's instruments and might not even know their true roles in its plans.

Although the Culture is a benevolent and ultra-liberal society, it often acts in destructive ways for what it sees as the greater good. Its tactics are pragmatic and sometimes they have to be ruthless, though they are never motivated by cruelty or a quest for military glory. All the Culture's actions lie under an ethical shadow, because there is always a question as to whether, all things considered, they can be justified. At one point in *Use of Weapons*, Zakalwe debates just such questions with Tsolderin Beychae, whom he wishes to recruit on behalf of the Culture. The goal is to oppose a group, the Humanists, who favor aggressive terraforming, oppose rights for sentient machines (proposing, instead, a sliding scale of rights for levels of sentience), and harbor imperialist ambitions. To the Culture, such views and the prospect of their implementation are anathema. But Beychae questions whether, or how far, the Culture's own moral standards are objectively justified, and how far it merely seeks to impose its own preferred practices on other civilizations.

Beychae is playing devil's advocate to an extent. He does, however, question whether the Culture's motives are pure and how far its practical and ethical judgment can be trusted. He points out that there are arguments for, as well as against, terraforming, and he suggests that the Culture's views on machine intelligence and intermixing of species go beyond cross-species tolerance to a more fanatical insistence that other societies take its approach of being run by machines and positively insisting upon intermixing of species. The Culture might appear morally superior overall to its various rivals, yet this debate is inconclusive.

As a further twist, it turns out that "Zakalwe" is not who he appears to be and that he has, himself, behaved monstrously in the past. If it comes to that, few characters in the Culture series are presented as morally clean. Those that come closest tend to be Artificial Intelligences rather than humans. At least this supports the Culture's positive attitude to sentient machines, but the Culture always seems slightly tainted—partly by its ruthless methods, and partly by a whiff of moral dogmatism. And yet, when is everything is thought, said, and weighed up, can the Culture ethically *not* interfere in pitiful situations? That's its condundrum.

Concluding Remarks

William Golding and Robert A. Heinlein could have written novels that did not employ science fictional elements such as a future war (in Golding's case) or an interstellar civilization (in Heinlein's). Without departing far from their intended themes, or even from the basics of the stories told in *Lord of the Flies* and *Tunnel in the Sky*, Golding and Heinlein could have used almost any scenario in which children or young people are stranded and isolated, and attempt to create a viable society. However, they drew on the icons and tropes of SF to make their scenarios more plausible and meaningful within the worlds of their novels.

Similar remarks could be made about many narratives, whether in the medium of print or film, or any other. Writers always have choices in how to explore their themes, including the ways in which they engage with philosophical and specifically ethical questions. Nobody absolutely must write science fiction. Nonetheless, the evolving SF mega-text is a bountiful resource for authors exercising their moral imaginations. It lends itself to an endless variety of themes, often in unexpected ways.

References

Blackford, R. (2016). *The mystery of moral authority*. Basingstoke, Hampshire: Palgrave Macmillan.

Blackford, R., & Broderick, D. (Eds.). (2017). *Philosophy's future: The problem of philosophical progress*. Hoboken, NJ: Wiley-Blackwell.

Broderick, D., & di Filippo, P. (2013). *Science fiction: The 101 best novels 1985–2010*. New York: NonStop Press.

Calhoun, C. (Ed.). (2002). *Dictionary of social sciences. Online version*. New York: Oxford University Press.

Churchland, P. S. (2011). *Braintrust: What neuroscience tells us about morality*. Princeton, NJ, and Oxford: Princeton University Press.

Gunn, J. (2006). *Inside science fiction* (2nd ed.). Lanham, MD: Scarecrow.

Jackson, P. T., & Heilman, J. (2008). Outside context problems: Liberalism and the other in the work of Iain M. Banks. In D. M. Hassler & C. Wilcox (Eds.), *New boundaries in political science fiction* (pp. 235–258). Columbia, SC: University of South Carolina Press.

Kant, I. (1991). *Groundwork of the metaphysic of morals* (H. J. Patton, Trans.). London, New York: Routledge (Orig. pub. 1785).

Lawler, D. L. (1977). *The Sirens of Titan*: Vonnegut's metaphysical shaggy-dog story. In J. Klinkovitz & D. L. Lawler (Eds.), *Vonnegut in America* (pp. 61–86). New York: Delacorte/Seymour Lawrence.

McGinn, C. (1997). *Ethics, evil, and fiction*. Oxford: Oxford University Press.

Russell, M. D. (n.d.). FAQ [for *The Sparrow*]. http://marydoriarussell.net/novels/the-sparrow/faq/

4

Future and Alien Moralities

Engaging with Ethical Theories

Though fictional narratives engage with ethical questions, they seldom comment directly on ethical systems such as Kantian deontology and utilitarianism. There are, however, some stories that go close, inviting us to judge the character of individuals or societies that seem to embody philosophical stances. Taken as a whole, science fiction appears somewhat hostile to utilitarianism, in particular, which is not to say that it offers a clear alternative. This is, perhaps, one point that arises from Iain M. Banks's Culture series. When the Culture interferes with oppressive societies, such as the warlike Idiran Empire in *Consider Phlebas*, its motives and actions may appear tainted, but the alternative is not obviously better. Hence, we cannot totally admire or totally condemn.

"The Ones Who Walk Away from Omelas" (1973) by Ursula K. Le Guin is one science fiction—or more likely fantasy—story that can be read as a direct critique of utilitarianism. I say "more likely fantasy" because this particular story could be science fiction, or fantasy, or neither. It offers no empirically plausible basis for the scenario it describes, but then again it is mainly a philosophical thought experiment rather than a narrative with a sequence of events. Le Guin invites us to consider a society where the happiness of the many is based on the misery of a single child, and thus her brief story reflects many earlier (and later) thought experiments involving immiserated scapegoats. "The Ones Who Walk Away from Omelas" suggests hostility toward utilitarianism or any similar consequentialist theory of ethics. This hostility is

© Springer International Publishing AG 2017
R. Blackford, *Science Fiction and the Moral Imagination*, Science and Fiction,
DOI 10.1007/978-3-319-61685-8_4

not linked to religious ideas (though Le Guin has shown an affinity for Taoism in much of her writing) or a Kantian analysis. The author's point is simple, but powerful: the unbearable knowledge of happiness based on another's undeserved pain.

A very different work, one that uses individual characters to represent positions in moral philosophy, is the comic-book limited series *Watchmen*, written by Alan Moore (1986–1987). In this series and the 2009 *Watchmen* movie (dir. Zack Snyder), a group of flawed superheroes play their roles in what almost amounts to a seminar in moral philosophy, though one involving an unusual degree of suffering and violence. Almost all of the supposed heroes are morally unattractive—if not outright repugnant—though most of them show at least some traits of character that we can approve. The nearest thing to the story's super*villain* is Adrian Veidt, known as Ozymandias, a businessman and scientist who fairly much embodies utilitarianism.

Against a backdrop of global tension that could lead to apocalyptic war, Ozymandias schemes to bring humanity together through an intervention that requires massive destruction and loss of life. In the original comic-book series, Ozymandias destroys much of New York, and kills millions of people, by unleashing a gigantic monster—thus faking an attack from space. In the 2009 movie, he inflicts nuclear devastation on New York and other major cities, faking this as an attack by the most powerful of the Watchmen, the godlike Dr. Manhattan. In both scenarios, Ozymandias is successful in presenting humanity with the appearance of a common enemy, and so averting what might have been Doomsday. It is left unclear, however, whether the resulting peace can last, since it depends on his plot not coming to light—and another character, the masked vigilante Rorschach, has already sent a journal containing his suspicions of Ozymandias to a right-wing fringe newspaper.

As the story ends, Ozymandias has prevailed—at least for now—and it is *possible* that he has saved a balance of many millions of lives. But his action carries an obvious taint. Then again, Rorschach, who might seem like the parody of a Kantian, with his uncompromising ethic of retribution and absolute right and wrong, is hardly a more attractive figure. Indeed, Rorschach's attempt to publish the truth might yet bring about more deaths on a colossal scale. Dr. Manhattan much resembles Winston Niles Rumfoord in Kurt Vonnegut's *The Sirens of Titan*. He ultimately supports Ozymandias, and even kills Rorschach, but (again like Rumfoord) he doesn't stick around in our local solar system, and while he knows much of the future he does not know the long-term result of Veidt's scheme.

Another character, the Comedian, is a coarse, cynical, unforgivably brutal egoist. Ozymandias kills him at the beginning of the action, but we learn much more about him through flashbacks. In all, the only characters who are largely sympathetic are those who seem most human and confused. Of these, Nite Owl and Silk Spectre are flawed (like everyone else in this story), but at least they come to love each other and they manage some ordinary decency in horrendously difficult circumstances. In the end, *Watchmen* does not offer us any characters or positions that are entirely admirable. Nite Owl and Silk Spectre are likeable, but even they seem rather tainted.

The Matrix and its sequels offer us yet another critique of utilitarian thinking. The protagonist, Neo, learns that the urban landscape where he has lived, grown, and worked is not the real world but a computer simulation of reality maintained by artificially intelligent machines. Neo is confronted with the choice of accepting ongoing life in his somewhat comfortable pseudo-reality—the eponymous Matrix—or discovering what he knows will be the unpleasant truth. He chooses the truth, which is revealed when he wakes up in a womb-like capsule on the surface of a ruined, scorched futuristic Earth run by the machines. *The Matrix* invites us to believe that Neo's choice to learn the truth was justified and worthwhile, though one of the more villainous characters, Cypher, regrets having made the same choice years before. Why not enjoy whatever pleasures are available, even at the expense of being cut off from reality?

The point here is not so much to advance a critique of consequentialist thinking as to ask what we really want from our lives. Is it happiness, perhaps in the form of sensual pleasures, or is it something rather different, such as living authentic lives, in the sense of lives in which we do our best to understand and face reality, however harsh it might turn out to be?

Deliberately or not, *The Matrix* recalls a passage in the philosopher Robert Nozick's *Anarchy, State, and Utopia* (1974). Employing what has become a much-cited thought experiment, Nozick suggests that we would not plug into a machine that could provide us with the mere simulation of experience in a maximally pleasant pseudo-reality. "Perhaps," Nozick writes, "what we desire is to live (an active verb) ourselves, in contact with reality" (1974, 45). Thus, Nozick challenges the overriding importance of subjective experience to how we ought to live. Many philosophers consider his argument to be a devastating critique of any utilitarian system of ethics based on maximizing pleasure or happiness. The thought experiment with the "experience machine" has, accordingly, pushed contemporary utilitarians toward the somewhat different concept of satisfying preferences.

Who's Afraid of the Brave New World?

Aldous Huxley's best-known novel, *Brave New World*, was first published in 1932, but it still speaks to us and provokes debate. Huxley intended it as a warning, and we are often confronted with dire predictions that we're headed down a slippery slope to the society that he depicts or something very like it. Perhaps most notoriously, Leon R. Kass, who chaired the President's Council on Bioethics during George W. Bush's US presidency, has argued on many occasions (e.g. Kass 2001) that we are headed in the direction of the Brave New World, since (as he sees it) much in modern Western society already resembles Huxley's scenario.

But why would this be a bad thing, even it were so? From the viewpoint of utilitarian ethical theory, Huxley's society and morality of the future have much to offer. Almost all of the characters in *Brave New World* are happy, or at least claim to be, so we might wonder why this is considered a dystopian novel. Bluntly, what is actually wrong with the Brave New World? It seems gentle, peaceful, and orderly. It is full of pleasures, with little in the way of pain, heartache, or worry. Kind treatment is given even to those malcontents who can't let go of values from less enlightened times: they are merely exiled to an island where they are allowed to do things their own way, for better or worse. If we project ourselves into this society, why do we like to think that we'd be among the malcontents?

Like *The Matrix*, then, *Brave New World* raises questions about what we really want from life. What does it mean for a human life to be "happy" or "flourishing" or to "go well"? Part of the problem lies in the promiscuous sex lives of the Brave New World's citizens, though deciding about this issue is not as straightforward as it evidently appears to Kass. In many ways, Kass is a traditional religious moralist, so it is unsurprising that he's so appalled. But it would seem superficial to condemn the Brave New World merely because it flouts the sexual *mores* of Huxley's society in the 1930s, or those defended by Kass, or even those commonly accepted in our own society. Later in this chapter, I will examine texts that take very different attitudes to sexual pleasure and expression.

The problem with sex in the Brave New World is not so much that there is a large amount of promiscuous and uncommitted copulation going on. Nor is it necessarily that these encounters take place outside the institution of marriage, or even that they happen outside relationships of romantic love. More disturbingly, sex in the Brave New World is reduced to much the same level as the many other amusing games that seem to dominate the characters' lives. In fact, we see no relationships of *any* kind—sexual or otherwise—based on deep, loving feelings. For the World Controllers several centuries hence, an inescapable emotional shallowness at all levels of society is a *good* thing. It removes a cause of frustration and dissatisfaction, and therefore of social instability.

The citizens of the Brave New World claim to be happy—but are their thoroughly planned and manipulated lives fortunate or flourishing ones? Reflection on a text such as *Brave New World* might suggest that more is required to flourish as a human being than subjective feelings of pleasure. It remains open to debate just what else is needed, but the lives of Huxley's characters should offer us some clues. At its root, the problem is not with the society's ubiquitous sexual promiscuity, the euphoric drug "soma" that its citizens enjoy, or its radical technology to control reproduction. Taken individually, these are symptoms of a deeper malaise. The problem is that all available forms of technology and social organization are employed in concert for a pervasive kind of control. Seen in this light, the Brave New World is a global society of willing, ignorant, cheerful slaves. They have pleasurable lives, but at the expense of losing anything remotely like real love, friendship, individuality, freedom, insight, or unique achievement. These are all suppressed through insidiously effective, though non-violent, means.

If this seems right, the crucial issue for our own society is not whether it displays superficial resemblances to the Brave New World. It is not whether we should permit (or encourage) recreational drugs, develop new reproductive technologies, or favor permissive approaches to sex. Rather, are we developing our own means to stunt and stifle human lives?

In his short monograph *Brave New World Revisited* (1958), Huxley offers his own thoughts on what is wrong with the Brave New World. He emphasizes the government's pervasive, systematic control of behavior. *Brave New World Revisited* is an urgent critique of social engineering and manipulation aimed at producing order and conformity. Huxley fears the emergence of a society that will stifle individual freedom, insight, initiative, creativity, and responsibility. He denounces catchwords and slogans, simplifications, suppression of unpopular ideas, and the repetitive dogmatism of propagandists. Instead, he proposes a political system where citizens are educated for self-governance. This would require teaching a few positive values, such as freedom, tolerance, compassion, mutual charity, and intelligence, but otherwise Huxley favors individuality and diversity in thought and ways of life.

Perhaps strangely, the utopian society of Pala depicted in Huxley's final novel, *Island* (1962), bears many points of superficial resemblance to the Brave New World. Like the Brave New World, Pala employs rituals, slogans, a powerful (freely available) drug, reproductive technology, and a degree of sexual permissiveness—yet these are employed in the service of producing free, mindful citizens and an ecologically sustainable set of institutions and practices. By contrast, the equivalent innovations in the Brave New World are the tools of a kindly totalitarianism, a regime of all-pervading paternalistic control. The Brave New World triumphs over whatever dissent it occasionally faces, but Pala is last seen going under the heels of power-hungry neighbors, high-ranked betrayers, and rapacious corporations. For Huxley, it seems, human flourishing and freedom are always fragile.

Brave New World is one of the great dystopian narratives of the twentieth century that reacted to the utopian visions of H.G. Wells in particular. Earlier in the century, E.M. Forster wrote an influential story, "The Machine Stops," first published in 1909. This presents an unattractive future in which human beings live underground in isolation from each other, and have become completely dependent on machines. This was one template for dystopians such as Huxley, and later for *The Matrix*. Ironically, however, Wells himself produced an even earlier template for dystopians with his novel *When the Sleeper Wakes*, first published in serial form 1898–1899.

Wells's Sleeper wakes, after two centuries, to a very altered London that has some initial attractions. He soon learns that his society's art forms and social *mores* have changed greatly, and not necessarily for the worse. The emphasis is on the sheer strangeness of the future society as experienced by the Sleeper, and until deep in the novel's action it remains unclear whether this future world should be seen as a utopia of sorts or as something far more troubling. The Sleeper is appalled and angered by the carefree attitudes to sex in the privileged levels of society where he mingles, but he seems to be more upset by the sheer scale of change that he must adapt to than by the unchastity of the new world with its discreet and socially valued class of prostitutes. This development can, of course, be seen as a prototype for the Brave New World's ubiquitous sexual promiscuity. At any rate, Wells increasingly shows the future world to be dystopian. In particular, the hedonistic lifestyle of the upper classes depends on the misery and toil of the masses.

Wells's descriptions of manual workers toiling in great underground factories under the watchful eyes of the Labour Police are reminiscent of *The Time Machine*, with its account of the Morlocks' origins in the industrial laboring classes. It is clear that the workers of London, who are differentiating into a distinct race, will evolve into the Morlocks (see Hillegas 1967, 47).[1] Wells suggests the beginning of a speciation event as a separate, internally

---

[1] At least, they are headed in that direction. Perhaps the events of *When the Sleeper Wakes* will forestall the eventual outcome seen in *The Time Machine*.

homogeneous, class of tired workers emerges, no longer mixing with other classes, and developing its own morality, physical features, and dialect. If this looks back to *The Time Machine*, it also presages later dystopian narratives, not least Fritz Lang's 1927 film *Metropolis*.

The clever dictator Ostrog, who comes to power during the action of *When the Sleeper Wakes*, presages similar figures in other dystopias. At one point he defends the new world during a conversation with the Sleeper, describing it as one in which wealth is power. This debate presages those between the John the Savage and Mustapha Mond in *Brave New World*, and between Winston Smith and O'Brien in George Orwell's *Nineteen Eighty-Four*. The Sleeper, who sees himself as uncivilized and likely to be shocked by new, and perhaps morally decadent, customs, may be Huxley's model for the Savage. The ubiquitous Babble Machines, which churn out state propaganda, are a forerunner for the propaganda efforts of the all-powerful states described by Huxley and Orwell. In all these narratives, the future is another place, but it's not one where we'd want to live.

The Ethic of Destiny

Utilitarianism receives a different critique from authors who favor what might be called the ethic of destiny. An early version of this within the canon of science fiction can be found in H.G. Wells's 1914 novel *The World Set Free*, which concludes by portraying the final days of a visionary, Marcus Karenin, who foresees humanity taking its own evolution in hand. Karenin imagines technologies that would grant total control over our bodies, abilities, and emotional responses. Many other authors depict similar striving for human improvement, whether through the outward expansion of civilization beyond Earth, the powerful enhancement of human abilities, or a combination of both.

Something of the kind can be seen in George Bernard Shaw's play sequence *Back to Methuselah*, which proclaims the need for greatly extended human longevity. *Back to Methuselah* almost certainly influenced the non-fiction writings of J.B.S. Haldane and J.D. Bernal (see especially Bernal 1970 [1929]) and the novels of Olaf Stapledon—particularly the latter's *Last and First Men* (1930) and its sequels *Last Men in London* (1932) and *Star Maker* (1937). A similar vision was given cinematic form in *Things to Come*, released in 1936, and based on texts by Wells. It is also common, even dominant, in the work of SF writers from the Campbell Golden Age, notably Isaac Asimov and Arthur C. Clarke.

This ethic need not be one of specifically *human* destiny. In considering the idea that we might one day be supplanted by intelligent machines, Asimov suggests that this might be an acceptable outcome:

> So it may be that although we will hate and fight the machines, we will be supplanted anyway, and rightly so, for the intelligent machines to which we will give birth may, better than we, carry on the striving toward the goal of understanding and using the Universe, climbing to heights we ourselves could never aspire to. (Asimov 1981, 163)

In Asimov's 1955 novel *The End of Eternity*, the powerful agency Eternity, which holds a monopoly on time travel and exists outside the ordinary course of time, interferes throughout human history. Eternity tweaks events discreetly but with the intent to create maximum benefit from its carefully chosen actions. It explicitly adopts a utilitarian moral philosophy—based on the adage of the greatest good of the greatest number—to guide its use of godlike power. Its operatives intervene surgically; they incrementally nudge historical societies into more and more peaceful forms and try to reduce the burden of human suffering. Eternity saves mankind from disasters such as global nuclear war, and it judiciously introduces medicines where they will produce most good without an overall adverse effect. Asimov controls the reader's attitude to Eternity, partly by showing its activities through the eyes of Andrew Harlan, a brilliant Technician who is capable of great insight but also misinterprets some of the crucial events going on around him. He is inclined to delusions of grandeur and shows a paranoid sensibility in which he is at the center of others' jealous plots. For all that, his grasp of the situation contains an element of distorted truth.

By the end, when it seems that Harlan has actually managed to destroy Eternity, we are led to see the organization as needed for humanity's survival and happiness. Harlan is persuaded to embark on one more mission to undo the damage he has caused. But in an additional twist, Eternity's interference is ultimately rejected in the novel's scheme of values. Harlan's lover from the far future, Noÿs Lambent, ensures that Eternity is destroyed after all. In addition, she makes a massive historical change of her own by ensuring that humanity obtains nuclear weapons in the twentieth century—far earlier than it would otherwise have done—thus risking human survival in the short term but opening up the possibility of expansion in space and the creation of a magnificent Galactic Empire.

As Lambent reveals, Eternity has altered what would otherwise have been humanity's destiny, creating in its place something safer, more restrained, and with less social and technological innovation. Through its incessant meddling, Eternity has radically delayed the discovery of an interstellar drive, and so held back humanity from exploring and colonizing the Galaxy. As a result, humanity's progress will be outstripped by aliens who will permanently confine our species to Earth. Humanity will die out, disheartened, over a further period of hundreds of millennia. To avoid this dismal outcome, Lambent engineers a literal end of Eternity.

C.S. Lewis was no utilitarian, but he reacted sharply against the ideas of progress and destiny that he found in the work of H.G. Wells, Olaf Stapledon, and J.B.S. Haldane. His 1938 novel, *Out of the Silent Planet*, and its sequels provided a riposte. When Lewis's protagonist, Arthur Ransom, gets mixed up with a couple of shady Wellsian scientists (one of them, Weston, was probably modeled on Haldane), he soon challenges them in moral debate. Lewis rejects any idea of sacrificing individuals for the sake of human progress, whether progress is pursued for its own sake or for utilitarian benefits, and he especially mocks the view that we ought to bend our efforts to assist the destiny of creatures, descended from humans, that might come to exist somewhere in the universe centuries in the future. Thus, Lewis deflated a moral position that was popular in the science fiction of the time and then continued as a thematic mainstay in much genre SF.

Lewis's novels also reject a wider, non-moral viewpoint that might seem implicit in modern science. As a genre, science fiction has tended to endorse the idea that science demystifies and disenchants the cosmos. By contrast, Lewis imagines a re-enchantment. He urges us to turn away from the mere numbers, as he sees them, that make humanity seem small, insignificant, and unexceptional in the vastness of space and depth of time. For Lewis, the universe remains pervaded by the presence of God and other supernatural intelligences. In *Out of the Silent Planet* and its sequels, he introduces beings that are explicitly the God, Satan, and angels of Christianity, with only the names changed (although *Out of the Silent Planet* does question exactly how neatly Christian angels line up with the *eldili* that Ransom meets on Mars).

Within Lewis's scheme of thought, the presence of these beings does not introduce anything whose actual existence he denies, and his books might be considered science fiction by an author who is theologically committed to the existence of supernatural entities. This further complicates the question of where we should place the borders for science fiction, which I discussed in Chapter 1. It would seem pedantic not to categorize at least *Out of the Silent Planet* as a science fiction novel. Its sequels—*Perelandra* (1943; also known as

Voyage to Venus) and *That Hideous Strength* (1945)—veer increasingly toward fantasy.

War and War's Alarms

Starship Troopers (1959) won Robert A. Heinlein a "best novel" Hugo Award in 1960. It is one of Heinlein's most loved works, yet it is also despised by many critics. Those critics notwithstanding, it is an important and thoughtful novel that merits close attention to its structure and implications. It is a key work not only in Heinlein's oeuvre, but also in science fiction's tradition as a literature of ideas.

Heinlein portrays an ongoing interstellar war between humankind and the alien "Bugs," but he does not show the outcome. The story is not, therefore, one of a military victory achieved against the odds, so what is *Starship Troopers* really about? The narrative voice gives us a clue: the story is narrated to us in the first person by the novel's protagonist, Juan Rico, who has clearly survived the events that he describes and is now reflecting—from a position of greater wisdom and self-understanding—on his life trajectory to date. Rico tells us of his formation as a military officer, from his confused decision to enlist through the accumulation of events that mold him into a man who can lead other fighting men. The essence of the novel is not humanity's war against the Bugs; rather, it is an individual's psychological (as well as physical) growth through hardship and dedication. As Rico undergoes training and combat, he comes to understand that military life is an honorable and (if things go well amidst the ever-present dangers) satisfying choice.

Rico describes events in a way that is slightly distanced from them: there is a certain amount of personal reflection woven through his narrative, and the prose is not as sensuous and textured as we might have found in a style of narration more focused on the protagonist's moment-by-moment thoughts and perceptions. Nonetheless, we are shown the hardships of military life in considerable detail. Indeed, Heinlein ups the ante by presenting us with a military system that is said to be extraordinarily efficient while also being very brutal by our current standards. It includes the routine use of flogging as a form of punishment.

In Rico's society, the right to vote is granted only to retired military personnel—not necessarily those who fought on the frontline, but anyone who volunteered for service and successfully served out their term. This limited franchise is not defended as a reward for virtue or as a recognition of intellectual superiority. Rather, the idea is that men and women who volunteered and served have thereby demonstrated their willingness to place the welfare of the many ahead of their own personal safety. Thus, they can be trusted with the vote. This system works well, we are told, in producing wise political decisions.

Since the action of *Starship Troopers* is set in the future, Heinlein can postulate new developments in human knowledge, not limited to technological advances. And so we learn that moral philosophy has become a mathematically exact science. From a philosopher's viewpoint, of course, this appears to be bluff. Despite much talk of mathematical demonstrations of ethical ideas, the classroom debates that Rico describes involve rather banal and sloppy arguments. Still, this narrative device enables Heinlein to suggest the possibility of a single true ethical theory (at least for human beings) based on various levels of survival: individual, familial, social, and so on, up to the level of our species itself. Just how all this is supposed to be reconciled is never explained, which is just as well in a book intended as a novel rather than an academic treatise. We are assured, however, that the future society can somehow justify all its political structures and decisions, including all the rituals, customs, and quirks of its military.

At any rate, Rico really does seem to mature as he accumulates experience; humanity's military forces really do seem to operate efficiently; and the regimen of harsh training and discipline evidently produces desired results. All of this could be criticized from various perspectives—not least one that emphasizes the novel's frequent sentimentalizing of violence—but *Starship Troopers* is, by and large, artistically successful on its own terms. At a minimum, it is a skilled exercise in recruitment propaganda: it shows military life as hard, rigorous, and unglamorous, and yet it gives *that* a sort of glamor of its own. By this, I mean that there is no glamor in the daily discipline and tasks of a working soldier, as presented by Heinlein. It is all acknowledged as difficult, tedious, demanding, and dangerous. Yet Heinlein builds a mystique around being the sort of person who would step up and embrace this way of life as a form of social responsibility.

No one could claim that *Starship Troopers* merely shows the honor and the plight of the troops. Instead, it openly embraces the necessity of war. At the same time, it displays pleasing literary skill. Even potentially embarrassing scenes, as when Rico's forty-something father follows his son into military service near the end of the book, are handled deftly.

Starship Troopers was rejected by Scribner's for its series of juveniles by Heinlein, and, indeed, despite its relatively young protagonist, it does not feel like a book written primarily for an adolescent audience. For younger readers, it offers a positive view of military life, but it is a novel of ideas—and primarily ideas in moral and political philosophy, at that—aimed at an audience of all ages. Whether or not we like its message and its explicit didacticism, it puts science fiction tropes to serious use in engaging with deep questions about the nature of a good society and a good life.

Joe Haldeman's *The Forever War* was published fifteen years later. It is the nearest thing within the science fiction field to a direct reply to *Starship Troopers*. At the same time, Haldeman's novel owes much to Heinlein's. It employs similar narrative structures and techniques. *The Forever War* insistently portrays the harshness of military training and the sheer terror, danger, and suffering of armed combat. We should, however, recall that *Starship Troopers* does not deny any of these or downplay them. They are always present in Heinlein's novel, and even emphasized. But Haldeman removes any sense of their mystique.

Like *Starship Troopers*, *The Forever War* is told in the first person by a narrator who has obviously survived the experiences that he recounts. Each novel portrays the military career of an individual soldier as he advances through the ranks. However, there is a different feel to Haldeman's prose: there is more focus on the grim details of training, work, and combat—as if we were there ourselves—and there is less reflection on the experience as a whole, viewed from the narrator's current perspective. Whenever reflections on war are offered, they come from a voice that sounds jaded and bitter. By contrast with Heinlein's finely honed officer, Juan Rico, Haldeman's protagonist, William Mandella, never comes to discern any benevolence or wisdom within the military forces that he serves. Very unlike Heinlein, Haldeman depicts high-level military commanders as ruthless in their exploitation of the men and women who serve under them. If *Starship Troopers* can be read as recruitment propaganda, *The Forever War* is pretty much the opposite—more like a warning.

Surprisingly, though, the eponymous "forever" war does *not* actually last forever. Unlike the war against the Bugs in *Starship Troopers*, it has an ending that is revealed before Haldeman's novel concludes. It turns out that the war was unnecessary, based on a misunderstanding for which human beings, and especially human military leaders, were mainly responsible. The dreaded and (to humans) visually disgusting enemy—the Taurans—were the relatively innocent party throughout, while humans were the aggressors. For Heinlein, war is inevitable and beyond moral criticism: its inevitability is an amoral feature of the background against which moral judgments are made. By contrast, Haldeman presents war not only as Hell (in fairness, Heinlein also does this, and almost as vividly) but as an especially futile, absurd kind of Hell.

Leaving aside its obvious critique of war and military organizations, *The Forever War* contains much in the way of social and moral commentary as it reveals the varied *mores* of several periods in future human history. Because Mandella spends much of his military service in space vessels traveling at relativistic velocities, he effectively keeps time-traveling into his future via time-dilation effects. That is, in fact, one reason why he renews his military service: not because he wishes to be alienated from human society but because, once he starts, there is little choice but to continue. There is no place for him in the radically changed society that he encounters the first time he returns to Earth from a battle with the Taurans. This aspect makes the war even harsher for personnel who serve in it.

Haldeman provides a happy ending in a way that Heinlein does not in *Starship Troopers*. The war is eventually resolved amicably and abandoned by both sides, and Mandella finally discovers a path out of military service, along with his lover, Marygay Porter. If we don't buy Heinlein's line about the necessity of war, we might actually see the narrative arc of his novel as more pessimistic than that of Haldeman's.

The science fiction genre has been used for a wide range of depictions of war. However, *Starship Troopers* and *The Forever War* are at opposite poles. One author (Heinlein) views war as an ineradicable feature of the human condition; the other (Haldeman) sees it as a form of madness that could be avoided with sufficient clarity and good will. Since *The Forever War*, Haldeman has continued to reflect on the cruelty and horror of war, and on why warfare persists. Most notable, perhaps, is *Forever Peace* (1997), in which the protagonists set out to pacify human nature once and for all.

Viewed as a whole, science fiction communicates ambivalent attitudes to war and warfare. Heinlein's views notwithstanding, there is often an overt anti-war message. In sub-genres such as space opera, however, any such message is undermined by the pleasure that we are clearly expected to take in spectacular battles, sometimes on inter-galactic scales. Popular science fiction in the cinema and on television relies heavily on the aesthetics of choreographed fight scenes. Still, we're supposed to blame the need for violence on whoever might be the bad guys.

In Wells's *The First Men in the Moon* (1901), the Grand Lunar—the ruler of the Selenites—is puzzled by many aspects of human institutions and conduct, but especially by the phenomenon of war. Unfortunately, the novel's central character, the eccentric inventor Mr. Cavor, explains human warfare to the Selenites with his usual naive, blustery gusto. It is after Cavor's descriptions, to the Grand Lunar, of the destructive power of late-nineteenth-century weaponry that his transmissions from the Moon to the Earth cease. By this point of the story, the Selenites have already witnessed how destructive men from Earth can be, with muscle and bone structure adapted for much stronger gravity than their own. The implication here is that Cavor has probably been killed by his hosts. At the least, they've prevented him from sending more messages back to Earth.

In George Orwell's *Nineteen Eighty-Four*, war is used in a different way. The three great powers of a near-future Earth, the superstates of Oceania, Eurasia, and Eastasia, maintain a permanent state of warfare among themselves, all in the interests of their respective ruling elites. This burns up surplus production, keeping most of the world's population in an optimally deprived and insecure state, despite the availability of resources that could provide economic security and leisure time for everyone. In other words, the superstates of the future conduct a perpetual war for entirely cynical purposes. Orwell views war with a cynicism of his own.

The Exotic and the Erotic

Science fiction depicts frontier societies—recently established societies developing their own values and *mores*—as well as much older societies located in remote settings. SF's frontier societies often begin as human colonies in locales such as the Moon or Mars, or sometimes in the asteroid belt. One classic example is the lunar society of Heinlein's *The Moon is a Harsh Mistress*, whose social arrangements and ways of doing things have diverged from Earth's. Likewise, Kim Stanley Robinson's mid-1990s Mars trilogy, consisting of *Red Mars*, *Green Mars*, and *Blue Mars*, shows Martian society diverging from its origins in the cultures of Earth after many years of colonization and innovation.

By the time we reach *Blue Mars*, which begins over a century after the colonizing expedition of the First Hundred early in the twenty-first century, things have changed greatly. We are shown a wide variety of new cultures arising on Mars, though one especially notable development is the rejection of patriarchal traditions and processes. Among the characters introduced in the final volume of the trilogy is the sensual, risk-taking Zo, who has rejected much of Earth culture, including those aspects brought to Mars by earlier colonists (although she is impressed by the deep-thinking and open-minded Sax Russell). Zo dies in a flying accident, perhaps casting some doubt on her risky lifestyle, and yet her life of intense challenges and pleasures is presented sympathetically.

Remote settings, whether in space or time or both, enable science fiction writers to present imaginary societies with unusual forms of social and family organization. These very distant societies can be employed thematically to issue dystopian warnings, offer utopian alternatives, or simply illustrate something of the range of possibilities outside whatever the audience assumes is normal. Distance in space or time also produces emotional distance, which enables a relaxation of traditional moral assumptions. Historically, this has granted science fiction (and modern fantasy) writers "a much wider latitude for commentary on sexual foibles" than that enjoyed by more mainstream authors (Brogert 1986, 87).

This latitude can be exercised for a range of purposes—often just to arouse the reader's (perhaps mild) erotic interest. But human sexuality is a fascinating theme in its own right, and sometimes SF engages with it seriously. There is also, I suspect, another aspect. It is difficult to imagine radically alien societies, and the result might leave the reader confused or unconvinced. But it is relatively easy to imagine societies much like the writer's own with just a small frisson of difference in sex-related customs. If the variations from the writer's society are limited to the socially approved amount of clothing, some local marriage practices, and perhaps some unusual attitudes to sex, this is not confronting for a moderately broad-minded reader. These superficial differences can function as a stand-in for more general or radical deviations from our own *mores*. For example, if the locals are much like us except for their habit of wearing nothing but jewelry, that can be a proxy for the *alienness* of an alien society—we're not in Kansas anymore!—but nothing important is at stake. No one's motivations need be much different from our own.

Western Christendom has a longstanding taboo against nudity, although this has decreased over the past half-century or so. Even before that, some licence was available to transgress the taboo in art and literature, especially if the subject matter was distanced in time or space, as with pictorial representations of scenes from pagan mythology. Nineteenth-century adventure novels, such as H. Rider Haggard's *She*, could get away with including naked or near-naked pagan women. Edgar Rice Burroughs portrayed Dejah Thoris in *A Princess of Mars* as naked and tantalizing, but the actual storylines for Burroughs' books set on Mars (or "Barsoom") were prudish by twenty-first century standards. A similar combination of nudity on alien worlds and rather decorous sexual attitudes can be found in E.E. "Doc" Smith's Skylark series.

The classic 1950s movie *Forbidden Planet* exemplifies the Arcadian and mildly erotic aspects of nudity in distant locations. Beautiful, young Altaira swims naked, wears scanty clothing, and keeps dangerous animals as pets. All this emphasizes the Edenic nature of her life, and her essential innocence—but also her potential for sexual experience. In this case, Altaira's sexual awakening is a significant plot point. By contrast with Altaira, who is a highly sympathetic character, some twentieth-century works of science fiction used nudity and eroticism to suggest a deeper moral decadence. For example, the False Maria robot—a villainous figure in Fritz Lang's *Metropolis*—is shown dancing almost naked. We have already encountered the dissipated upper classes in Wells's *When the Sleeper Wakes* and Huxley's unfavorable depiction of a sexually promiscuous society in *Brave New World*.

Magazine covers from the early decades of genre science fiction often portrayed scantily-clad women, but this was seldom emphasized in the stories themselves. Indeed, the specialized SF magazines of the Gernsback era and the Golden Age were rather asexual if we ignore their covers. More realistic depictions of sexuality did appear in some science fiction, but this was outside the magazines. For example, Philip Wylie dealt quite realistically with sex in his influential novel *Gladiator*, published in 1930. But science fiction written for a dedicated audience did not include frank sexuality in its mix.

During the 1950s, this began to change. Philip José Farmer's story "The Lovers" (1952) introduced a sexual relationship between a human and an extraterrestrial, opening up a theme that became more common in the decades that followed. Farmer later developed the idea further in a novel-length version, *The Lovers* (1961). The protagonist's lover, Jeanette, turns out to be an insect-like creature that can imitate the female human form and thus insinuate its way into human social groups. Prior to the arrival of humans from Earth, such creatures imitated one of the two major intelligent species on the planet Ozagen—a species very like Earth humans. When the protagonist, Hal Yarrow, inadvertently brings about Jeanette's death—by attempting to wean her off her imagined alcoholism—it is depicted as a tragic event even though Jeanette was neither human nor one of Ozagen's indigenous "humans." *The Lovers* contains much satire aimed at colonialism, theocracy, and anti-sex forms of asceticism. Ultimately, it takes the side of Ozagen's native species against the humans who plan to take over their world.

The 1960s was the era of sexual revolution, and science fiction played its part, especially with Heinlein's *Stranger in a Strange Land*, first published in 1961. For a generation of SF readers who grew up in the 1960s and 1970s, this was a book that changed lives: a huge, bizarre, magical, loosely-knit satire of nearly everything. It soon achieved cult status, with an audience far beyond science fiction's usual reach at the time. It celebrated the human body (the characters seem to be nude as often as not), advocated open sexual relationships, and laughed at politics, jealousy, organized religion and moral convention. It fitted neatly into the hippie ethos of the later 1960s—it was embraced as a kind of gospel of free love—and it has never entirely lost a readership.

A much longer "original uncut" version was published in 1991, and fans now compare the merits of this version and the relatively taut, if sometimes more obscure, version from the 1960s.[2]

Stranger in a Strange Land recounts the adventures of Valentine Michael Smith (or Mike), a young man who was born on Mars and raised by the Martians, before being brought to a near-future Earth. He is befriended by wise old Jubal Harshaw, who becomes his mentor and protector—then eventually something more like a disciple. Mike becomes a messianic figure, and he is eventually murdered by an angry mob. Throughout, Heinlein rather hammers the point that there is nothing privileged about the moral system of any one culture. Jubal insists on this again and again in his lengthy dialogues with other characters, notably Ben Caxton (a reporter) and Duke (Jubal's handyman). Jubal presses on the hapless Duke his claim that there is no quality of instinctive knowledge even in modern Western society's heavily moralized aversion to cannibalism. This view receives its ultimate validation after Mike is killed, when his followers—including Jubal and Duke—end up eating his flesh in a tasty broth.

The novel's detractors refer to its apparent diffuseness and other oddities of structure and style, but much of this can be justified. Importantly, *Stranger in a Strange Land* should not be read as a traditional novel with tight plotting, realistic characters, and lifelike dialogue. As I expressed the point elsewhere, "*Stranger* is more like a Menippean satire than a psychological novel: its amusement lies very much in its snippets of outlandish news reports, bad verse, Socratic dialogue, fable and philosophy" (Blackford 1985, 67). It is all held together by a loose, though clearly apparent, overall structure, by a pattern of dominant references and images, and by an obsessive interest in various topics that are satirized. The pleasure of the text comes largely from its seeming digressions.

Though *Stranger in a Strange Land* does not take us on any convincing psychological journeys, it is sometimes emotionally moving. This requires at least some engagement with its main characters, especially Mike and Jubal, and that, in turn relies on a degree of psychological consistency. As readers, we cannot be engaged by characters who are psychologically capable of *just anything*. A full account of the novel's unusual form of coherence would explicate Mike's role as a Christ figure of a rather pagan kind, one who brings salvation through self-perfection rather than through submission to an external God or any metaphysical force.

---

[2] Speaking for myself, I wish that Heinlein had been able to produce a definitive author's cut with the strongest material selected from both versions.

Stranger in a Strange Land does have weaknesses, especially when read many years after its initial publication. Not least, there is sometimes an uncomfortably sexist feel to it. Perhaps this is more troubling in such a plainly didactic book than in a more straightforward narrative that carries its author's opinions more lightly. Whatever might have been Heinlein's intention, Jubal's three secretaries, Anne, Miriam, and Dorcas, are scarcely differentiated through much of the narrative, and they seem interchangeable for most purposes. And yet, it would take a certain narrowness of literary sensibility not to enjoy *Stranger in a Strange Land*. At the least, it's a joyous satirical romp, and its unashamedly sex-positive message remains refreshing all these years later.

The polymorphous and polyamorous arrangements of Mike's "Nest" of companions—with free love among male and female "water brothers," but the exclusion of outsiders—are not intended, we might presume, as a literal social blueprint. Rather, we are left to make up our own minds as to how the sexual and other arrangements in our society could be improved. Heinlein conveys an overall sense that the Nest would be an improvement on what we have (or at least what we had in 1961), even if nothing like it could be adopted *holus bolus* in the absence of an alien miracle worker like Mike.

By the end of the novel's action, the Nest is well on its way to becoming the dominant church of a new religious era, spreading an attitude of sexual celebration and a doctrine that equates awareness—preeminently intelligence—with God. Perhaps more disconcertingly, we find Mike cheerfully discorporating (obliterating the bodies of) very large numbers of individuals who become his enemies through the sequence of events, or who are simply of vicious character. This makes sense within a fictional world where all human beings are immortal and are merely sent back to begin again when they are killed. However, similar attitudes can be found in other stories and novels by Heinlein, including *Friday* (1982). Again, the main character (a genetically engineered "artificial person") does not hesitate to act with deadly violence when it seems needed.

Much of Heinlein's work seems to promote an ethic in which the superior—we might, in a secular sense, say Elect—people are justified in using all levels of violence required to protect themselves and each other against the rest. While the characteristics of the Elect include a default approach of non-violence (along with intelligence, physical beauty, generosity, non-conformity, and sexual openness!) they do not include any inhibition about killing when it seems indicated by the circumstances. For better or worse, novels such as *Stranger in a Strange Land* and *Friday* imply an ethic radically different from the morality in traditional Christian teachings.

During the following decades, science fiction began to deal with sexual themes with a vengeance. Because it examines social changes, the SF genre provides an ongoing literary forum for such issues as whether monogamy should be retained. It sometimes critiques monogamy from an alien viewpoint, as in *Stranger in a Strange Land*, or describes alternatives, such as the line marriages of *The Moon is a Harsh Mistress*. In Arthur C. Clark's *Imperial Earth* (1975), exclusive marriage and sexual possessiveness are said to be things of the past, and Clarke shows us successful polygynous and polyandrous marriages in his hard SF masterpiece *Rendezvous with Rama*.

Some science fiction works describe unusual familial/sexual arrangements for their alien characters, as in Isaac Asimov's *The Gods Themselves* (1972), without considering anything very radical for future humans. However, Samuel R. Delany, Philip José Farmer (of course), John Varley, and others imagined more extraordinary sexual scenarios, some of them involving the interactions of humans and non-humanoid aliens.

Science Fiction's Critique of Gender Roles

Inevitably, many science fiction writers have challenged the traditional roles assigned to men and women, respectively, and to question the expected relations of the sexes. Some writers have depicted feminist utopias: societies where, broadly speaking, feminist goals have been obtained. Others have imagined dystopias—depictions of societies in which women are oppressed, as with Margaret Atwood's *The Handmaid's Tale*. These dystopian stories can issue warnings and/or critique sexism in the author's own society.

One early feminist utopia was Charlotte Perkins Gilman's *Herland*, originally published in serial form in 1915 and for many years inaccessible to most readers. It was republished in 1979—at the time of a flurry of feminist utopias—and is now regarded as a classic of its kind. In *Herland*, the utopian society is a variant on the idea, popular at the turn of the twentieth century, of a lost world. The narrative begins with the journey by a small group of men to an isolated land populated only by women, who reproduce by parthenogenesis. They discover that women are perfectly competent at tasks traditionally assigned to men in the author's own culture. More than that, women have built a working society with no experience of social subordination or war.

The idea of a society of women was not completely new, but it was unusual in 1915. Since the 1970s, it has become routine in feminist SF, perhaps even passé. During the 1970s, feminist utopias became commonplace, prompted by the ferment of ideas swirling around the Women's Liberation Movement of the 1960s and early 1970s. This is often referred to today as "second-wave feminism."[3]

Second-wave feminism did not arise in a vacuum. It reflected social changes at the time, including the changing needs of capitalist labor markets. Accordingly, some proto-feminist science fiction novels preceded the first manifestos of second-wave feminism. Philip Wylie's *The Disappearance* (1951) was one notable proto-feminist work from a male author. As a result of some cosmic glitch, the world's men and women awaken in two separate timelines or universes, one for each sex. This provides Wylie with an enabling device to examine homosexuality, and it especially permits him to explore the respective natures of men and women. As events turn out, men make a mess of their world while women are better off without men around. The timeline in which women fare well without men can be seen, therefore, as a feminist or proto-feminist utopia. Theodore Sturgeon's *Venus Plus X* (1960) was another proto-feminist novel by a man—and, unsurprisingly for its time, it contains allusions to *The Disappearance*. The critic Judith Brogert mentions Sturgeon's presentation of "the damaging effects of traditional sex roles" and rejection of "both matriarchal and patriarchal societies" (1986, 94).

Venus Plus X employs two alternating strands of narrative that never join up, although they reflect each other thematically. One strand involves a suburban couple, Herb and Jeanette Raile, who see themselves as having transcended gender-based work and parental roles, and as forming an equal partnership. Though Herb and Jeanette are endearing individuals, they also show an obvious measure of hypocrisy and lack of self-awareness. In particular, they repeatedly fall back into training their children in conventional gender roles and sexual *mores*. Throughout the novel, there is much satire of their clueless attempts to be modern (as they think of it) about gender and sexuality. Herb worries about how his daughter will be treated as she grows up: he does not want her to be discriminated against for her sex. But at the same time, he worries that his society is becoming too androgynous. He appears well-intentioned but plainly confused.

The science fictional strand of the narrative involves the adventures of Charlie Johns, who has been snatched by a mysterious hermaphroditic civilization (the Ledom), apparently from the future. Much of the story consists of his being shown, in conventional literary utopian fashion, how this society works and how everything about it is better than in his own society. The Ledom have advanced technology in the form of the A-field ("A" for analog),

---

[3] It is thus distinguished from still later "waves" of feminist thought, controversy, and activism.

which can be used for a wide range of effects, and the cerebrostyle, which allows for almost instantaneous learning. Their hermaphroditism is initially explained to Charlie as a mutation that has turned out to be superior, ridding the society of sex-based subordination—and along with this, they seem to have avoided other kinds of domination and social subordination.

However, Charlie is disgusted when he eventually learns that the Ledom have not whisked him into the future, and that they are not a naturally occurring mutation; rather, they are the product of biomedical manipulations. Having been favorably disposed to the Ledom up to this point, he rebels because he sees them and their society as transgressions against nature.

As events turn out, Charlie is being used by the Ledom to gauge the reactions of human beings. They want to test whether the time is yet ripe for them to reveal themselves. Charlie's revulsion when he's told the truth makes clear that the time has not yet arrived: human civilizations are not ready. As a further twist, it turns out that "Charlie" is not even the real Charlie. The original Charlie died in a plane crash, and the Ledom took a cerebrostyle record of his mind and memories from an earlier time in his life. As for the Ledom, their society does not exist as a utopia intended to supersede human societies. They were created by a group of scientists, who also gave them the technology of the A-field and the cerebrostyle, to study and protect humanity until it is ready to function without its violence and oppressive systems. As *Venus Plus X* concludes, it appears that a nuclear war has started among humanity, though the Ledom possess technology to shield themselves from the fallout. They are prepared to wait, no matter how many times humanity destroys itself.

The feminist wave of the 1960s and 1970s soon brought a wave of overtly feminist science fiction. Important novels of the time include the four volumes of the Holdfast Chronicles by Suzy McKee Charnas, beginning with *Walk to the End of the World* (1974), *The Female Man* by Joanna Russ, *Woman on the Edge of Time* by Marge Piercy, *The Wanderground* by Sally Miller Gearheart (1979), *Native Tongue* by Suzette Hayden Elgin (1984), and *The Gate to Women's Country* by Sheri S. Tepper (1988).

Though they vary considerably among themselves, many of these novels include societies without men, invariably suggesting that they would be improvements on current and historical societies. *The Female Man* involves interactions among women from four different realities. Joanna's world resembles Earth in the 1970s, with the stirrings of a feminist movement to challenge gender roles and women's subordination. Jeannine lives in a dystopian world in which the Great Depression never ended—and in this world there is no sign of a feminist movement. Jael is from a world in which societies of men and women are literally at war, while Janet lives in the all-female society of Whileaway. Whileaway is the nearest thing in the book to a feminist utopia, all of its men having died, or perhaps been killed, 800 years before. As it turns out, Jael is the instigator of events, seeking cooperation from the other women in a struggle against men.

Woman on the Edge of Time portrays the functioning of a future utopia—the community of Mattapoisett—with people of both sexes. Here, women have achieved freedom from male domination, and sexual interaction is free of guilt (though not of all interpersonal conflict). Ectogenesis is used to separate reproduction from recreational "coupling," and there is no practice of marriage. Individuals may have unlimited numbers of "sweet friends," while three social mothers (who may be of either sex) are provided for each child. Piercy also subverts the importance of gender roles by introducing the unisex pronoun "per" as an alternative to gender-specific nouns.

Ursula K. Le Guin's *The Dispossessed*, published in 1974, depicts a similarly utopian, though degenerating, society. Once again, many forms of sexual involvement are acceptable, but the focus of interest and sympathy is on monogamous, heterosexual relationships. Indeed, one of the problems that has arisen in Anarresti society is its failure to consider the interests of closely bonded couples. At this point, mention should also be made of Le Guin's earlier novel, *The Left Hand of Darkness*, which presents the planet Gethen, whose people do not have a fixed biological sex. It makes no sense on Gethen for there to be established gender roles, though this does not entail a utopian world so much as a recognizably human world without the institution of women's subordination.

By contrast to feminist utopianism, Margaret Atwood's *The Handmaid's Tale* is in the dystopian tradition of Huxley and Orwell. It is set in the Republic of Gilead, a near-future society resulting from the takeover of the US by a movement of Christian nationalists who call themselves "the Sons of Jacob." Within the resulting theocratic state, women are subordinated to men to an extreme degree, justified by passages in the Old Testament (the Hebrew Bible). Women have almost no rights or freedoms, and are regarded by the state as objects to be used for reproduction.

Atwood's scenario includes a shrinking human population, partly from widespread use of birth control techniques, including abortion, but also from a physical environment poisoned by nuclear leakage and other toxic emissions and waste. In response, the wealthy men of New Gilead have fertile young women assigned to them as "handmaids": more bluntly, these women are sex slaves. They are kept in addition to wives, and a handmaid is expected to bear children on behalf of the wife. The man, referred to as a Commander, has sex with his handmaid in a ceremonial process that also involves his wife. The main narrative is told from the viewpoint of a handmaid who is known only as "Offred," signifying that she is the possession of Fred, the particular Commander to whom she has been assigned.

The Handmaid's Tale gained its plausibility from the theocratic and patriarchal component of twentieth-century (and indeed, twenty-first century) right-wing politics in the US. The early 1980s, when the book was written, corresponded with Ronald Reagan's presidency, anxiety over the AIDS epidemic, the resurgence of conservative Christianity as a political force, and the rise of some groups not all that different from the Sons of Jacob. It made sense then—and perhaps still—to perceive women's basic rights as fragile, even in Western societies, and always at risk of being overturned by unforeseen social shocks.

But Atwood ultimately provides a note of hope. As the book concludes, Offred's story is framed as having happened about two hundred years earlier than an academic conference that we are shown taking place in June 2195. The novel's main text is revealed as a scholarly reconstruction, based on tapes that cannot have been made within the previous 150 years. Although the tapes' exact provenance is unknown, their basic account is regarded by conference goers as authentic. From the perspective of these comfortable scholars, the theocracy of Gilead has long vanished and a very different society has now emerged. This narrative device resembles (and may have been inspired by) one used in Orwell's *Nineteen Eighty-Four*. That equally bleak novel concludes with a document examining Oceania's artificial language, Newspeak, in a way that suggests the repressive superstate of Oceania no longer exists.

The scholars from 2195 look back on the Republic of Gilead with a certain detachment, even to the extent of making cold, sexist jokes about Gileadean society. In a strikingly obtuse lecture, Professor James Darcy Pieixoto goes out of his way to counsel his audience not to pass moral judgment on Gilead, since it was under specific pressures, and also, he claims, because such judgments are culture-specific. Atwood presents Pieixoto as a buffoon, and the overall impact of the novel is to repudiate such smug relativism. For the well-fed conference goers, of course, Gilead lies in the past where it represents no threat. They are more worried about missing out on lunch if question time goes too long. By implication, our own reaction should be very different.

The Handmaid's Tale blames Gilead's system of institutionalized misogyny on underlying currents within American society, the use (or misuse) of sociobiological theory relating to sex differences, and, most obviously, fundamentalist varieties of Christianity. More unusually for an overtly feminist novel, it also appears to place some blame on trends within feminism itself.

The early 1980s was a period of sharp divisions in the feminist movement, following a change in its motivations and emphases during the later 1970s. There was an increased priority placed on women's safety, as opposed to women's liberation from constraining laws and social *mores*. When feminist movements emphasize the vulnerability of women, Atwood implies, they play into the hands of theocrats.

Explicit feminism aside, fictional narratives challenge conventional gender roles whenever women are depicted as acting competently in traditionally male roles and occupations. The implied feminist statement is augmented when these depictions are given a degree of verisimilitude. It may be somewhat undermined to the extent that competent female characters are served up for the viewing pleasure of heterosexual men. Many superheroines and "woman warriors" are presented as hypersexualized and coded as sexually available. The artistic legitimacy of such characters is an issue of ongoing controversy, and perhaps they do have a worthwhile cultural role. But the feminist implication seems clearer when competent female warriors are *not* strongly sexualized, as with Imperator Furiosa in *Mad Max: Fury Road* (dir. George Miller, 2015) and Jyn Erso in *Rogue One: A Star Wars Story*. Such characters are often welcomed by feminist commentators.

Outside of such action-adventure scenarios as the Mad Max and Star Wars franchises, science fiction has played a valuable social role by providing believable representations of female scientists, astronauts, pilots, and other highly trained professionals. In discussing Robert A. Heinlein's *Stranger in a Strange Land* earlier in this chapter, I mentioned a degree of sexism that shows through for a contemporary audience. In fairness, Heinlein was also an innovator in depicting competent women in supposedly male occupations. Indeed, as I mentioned in Chapter 3, the female characters of Heinlein's *Tunnel in the Sky* seem more level-headed and practical than their male counterparts.

Delany's *Stars in My Pocket Like Grains of Sand*: Science Fiction as Subversion

Samuel R. Delany's *Stars in My Pocket like Grains of Sand* is a helpful case study, showing some of the limits to which science fiction can be pushed. Although it was first published in 1984, over three decades ago, it remains at the cutting edge in its subversion of current social and moral assumptions. This novel is actually two stories, or perhaps even three. The first is an almost self-contained novella, listed as the Prologue and entitled "A World Apart." It is followed by a set of "Monologues," with a final "Epilogue."

"A World Apart" tells the story of an (initially) unnamed slave on an unnamed world. Later, we find out that his name is Korga—he comes to be called "Rat Korga"—and his planet is Rhyonon. Korga is a misfit on a very backward planet. On the first page we are introduced to him at an institution that uses a process called "Radical Anxiety Termination." This involves destroying certain neural pathways in the brain so as to turn off the capacities for aggression, anxiety, and original thought. It is thus a kind of futuristic, and more drastic, equivalent of lobotomy: once subjected to the RAT treatment, a "rat" is completely tractable, even to the extent of being willing to sleep in his own excrement.

The social practices on Rhyonon declare that rats should be made the slaves of institutions that need cheap menial labor. The polar research institution where Korga spends most of his time withholds from him even the most universal and fundamental dignities. He lives a life that appears fantastically debased and humiliating, and there is nothing to encourage us to see it in any other way. However, the novel goes on to present us with a series of apparently degraded lifestyles, actions, and desires, and leaves us to sort them out in their value and their ethical acceptability.

The main action of "A World Apart" involves a woman who seeks the company of her own slave; she is a sadist who wants someone for sexual use and abuse. In a key episode, she demands that Korga have sex with her (he is not interested since he is entirely homosexual, but his RAT treatment does not let him refuse), to permit her to whip him bloodily, to permit her, too, to spit on him without defending himself. But she gives him a glove-like device that plugs into his neural circuits with the effect of healing his mind, while, at the same time, she grants him the release of knowledge. We learn that, under certain conditions, a rat can absorb information many orders of magnitude faster than somebody who has not had the RAT treatment. The woman wants to be able to spit upon a man who has read all the books she has not, so she sets him to read a collection of literary works through direct neural input, gulping down the equivalent of tens of thousands of pages within seconds.

The description of Korga's first experience of reading—and, with it, an exponential expansion of his consciousness—is impressive as a straightforward science fictional rendering of experience beyond the edges of anything we could ever encounter. Delany sensitively portrays a feeling of splendor, joy, and completeness. This is the other side to *Stars in My Pocket Like Grains of Sand*: although it presents degradation, it also shows startling forms of splendor. More radically, it seems to challenge the distinction between them.

Exactly how and where we make that distinction might, the book seems to insinuate, depend on what cultural assumptions we bring.

In the case of Rat Korga, the woman who sadistically degrades him is the same woman who makes available to him the vistas and the splendor of thought and intellect. When he is eventually "rescued" from her, we feel immediately that he has suffered an overall loss. Indeed, we are given a heartbreaking description of what Korga is like thereafter, wistfully and pathetically teaching his fellow rats to wear one work glove, in memory of the device that briefly gave him his mind and a mental world—before it was torn from him, and both mind and world rushed away.

The scene with the glove-device contains some trickery, and even jokes. It turns out that Korga dipped into a pile of women's literature by mistake, and he failed to realize that what he took to be the literary canon of his world was more like a literary ghetto. (There is also a hint of self-deprecation here, since Delany is a writer whom we'd expect to champion alternative literary canons.)

At the end of the Prologue, Korga's planet is destroyed, and with it, apparently, Korga himself. In the novel's larger story, however, it becomes clear that Korga is possibly the only survivor of the cataclysm. This larger story is narrated by Marq Dyeth, a small bearded male human being. Rat Korga and Marq Dyeth are spectacularly sexually attracted to each other, and from this point *Stars in My Pocket like Grains of Sand* is mainly concerned with their lusty relationship in the context of immense political conspiracies that affect the lives of humans and other intelligent beings on six thousand planets.

Part of the difficulty that we might experience in reading this novel is that it takes away many of the codes that we habitually employ to construct pictures of characters and their actions. Most obviously, it dispenses with the simple distinction in language between male and female people. The word "man" for "male human being" is said to be an archaism seldom encountered, while the pronouns "he" and "she," and related words, are not used to denote biological sex or even gender identity. Rather, they are used to distinguish between the mass of humanity (together with other intelligent species) and those individuals by whom the speaker is sexually excited. In Marq's universe, intelligent life-forms are called "women," and the pronouns "she" and "her" are standardly used. The pronouns "he" and "him," by contrast, indicate that the speaker refers to someone who sexually excites her.

Because of the differentiated roles and values traditionally assigned to men and women in Western society, we might too often tacitly presume that characters in books (and other human agents whose doings we hear about) are male unless we have evidence to the contrary. Although this is under challenge, it is reinforced by traditional English usage, which has long employed the masculine gender in many contexts that include girls and women. In *Stars in My Pocket Like Grains of Sand*, by contrast, we pervasively encounter what strike us—though not Delany's characters—as feminine pronouns. The effect seems to corroborate the idea that language shapes social assumptions. If we catch on quickly, though, we can get used to the idea that our "masculine" pronouns signal the sexual excitement aroused in the speaker by the person spoken of.

Delany implies that language, as well as shaping and reinforcing cultural attitudes, also manifests them. In the highly permissive universal meta-culture that he has created for *Stars in My Pocket Like Grains of Sand*, it is not merely *acceptable* to reveal openly when one is sexually excited and by whom. Such revelations are *forced* by the language itself. It as natural and inevitable for people to distinguish, in the course of speech, who sexually excites them as it is for us to distinguish whether a person being spoken about is male or female.

Apart from its innovative language, *Stars in My Pocket like Grains of Sand* challenges our assumptions as to what sort of descriptive details should be selected to evoke people. If we want to sort the characters as male or female, Marq Dyeth's descriptions are little more helpful than his pronouns. In selecting evocative details, he will often refer to veins, scars, calluses, and fingertips, rather than, for example, hair, breasts, and facial features. Marq (and hence Delany) provides as much sensory detail as the average writer, but not necessarily the kind we are used to. This draws attention to the way we depend on being told *certain kinds* of detail in constructing pictures and assigning identities. It also, of course, adds to the sense that we are in somebody else's very different mental world.

At the thematic center of *Stars in My Pocket Like Grains of Sand* is an idea of degradation, debasement, or abjection. This takes many forms: a great deal of emphasis is placed upon nudity—which can conventionally signify a lowering of status combined with great vulnerability. Stripping a victim or enemy is a method of humiliation, attacking some sources of identity and pride. At the station where Korga works in the early part of the book, he is treated negligently and demeaningly in that he is not adequately fed, is not provided with toilet facilities, and is forced to work naked. However, nudity can have other cultural meanings, up to and including godlike transcendence of human limitations. In *Stars in My Pocket Like Grains of Sand*, nudity is often presented positively, as it is in Heinlein's *Stranger in a Strange Land* and many other science fiction narratives. Delany's characters are shown as going naked at Dyethshome, where Marq and his fellows live in symbiosis and sexual interrelationship with the evelmi, a race of six-legged reptiloid aliens.

As already touched upon, some elements of *Stars in My Pocket Like Grains of Sand* simultaneously convey ideas of degradation and splendor—or transcendence. Many of the experiences described, although emotionally potent, would seem ethically neutral upon dispassionate analysis. For example, it is hard to believe that anyone should feel moralized concern about Marq's sexual interest in calluses and bitten fingernails. But some aspects of his way of life, such as his sexual involvement with reptiloid aliens, would be regarded by conventional moral wisdom as both disgusting and wicked. Delany is provoking us: he challenges our assumptions of what experiences are degrading, repugnant, immoral, or shameful.

Marq Dyeth is employed as an industrial diplomat, someone who deals with trading relationships among many cultures and even species. For him, it is second nature to assume that responses to situations, together with the very concepts that are employed in his interactions with others, vary fundamentally from culture to culture. He has internalized this assumption so much that, ironically, he sometimes appears bemused when not all others feel likewise. In these contexts, his bemusement is a civilized emotional response. Perhaps an outer limit of acceptable behavior is established when he encounters the proclivities of a male sadist, Clym, early in his narrative. Clym, we learn, is a professional psychopathic killer, as well as being sexually excited by torture. Even in this extreme situation, Marq declines Clym's advances coolly enough by our standards, and the narrative emphasizes his sudden sexual distaste rather than any *moral* repugnance. In fact, the tone of the passage is comic self deprecation (at Marq's inability to cope with such an experience as meeting Clym) rather anything else.

Concluding Remarks

The evolving science fiction mega-text has absorbed ideas that arose during the social transformations of the 1960s and 1970s. Outside the field of science fiction, the transformative agenda of the sixties and seventies was soon opposed by a range of broadly reactionary movements, including the Christian conservatism that reasserted itself at the end of the 1970s. These, however, obtained little foothold in genre science fiction or in contemporary art and entertainment more widely. Science fiction thus tends to present broadly "progressive" visions of the human future. It does not, however, follow that the SF field is politically united. It does include viewpoint diversity, but its allegedly right-wing practitioners are more likely to be libertarians of one sort or another than traditionalists or religious conservatives.

Open portrayal of sexuality and candid examination of sexual themes are now commonplace in the SF field, and some writers present sexually libertarian and polymorphous futures. As far as I'm aware, and notwithstanding Delany's *Stars in My Pocket like Grains of Sand* and a few other exceptional texts, inter-species sex involving humans and non-humanoid aliens remains somewhat taboo in science fiction. However, inter-species sex involving more humanoid characters has become routine. Many years after Philip José Farmer stirred things up with "The Lovers," an inter-species love affair was crucial to the plot of *Avatar*, the blockbuster movie directed by James Cameron (2009). As events unfold in *Avatar*, the human protagonist falls in love with a beautiful—albeit huge and blue-skinned—woman from the alien Na'vi.

That said, a series of events in the late 1970s and through the 1980s undermined the sexually libertarian aspect of the sixties-and-seventies agenda. These events included, perhaps, a degree of social fatigue with sexually libertarian ideas, a changed emphasis within feminism in the late seventies,[4] the aforementioned resurgence of religious conservatism, and the AIDS crisis that first came to notice in 1981. This degree of retreat is also reflected in the SF field. As a result, the more taboo-busting works of Farmer, Delany, Varley, and others now have a somewhat dated look. Thus, *Stars in My Pocket Like Grains of Sand*—from the mid-1980s—represents a high-water mark in sexual taboo-busting from SF authors with serious literary credentials.

Science fiction has readily absorbed feminist ideas and continues to absorb ideas relating more generally to the rights of historically oppressed groups. The idea of single-sex societies no longer appears confronting, but this also means it has lost its emotional impact. The once-unusual device of language without gendered pronouns is likewise no longer so unusual: it has become a familiar trope in the SF mega-text. For example, in Ann Leckie's Imperial Radch series, commencing with *Ancillary Justice*, the language used by the Radch Empire does not assign sex or gender. Leckie renders this by using female pronouns throughout unless a different language is being used by the characters. Like Delany in *Stars in My Pocket Like Grains of Sand*, Leckie forces us to get by without worrying too much about characters' sexual characteristics.

After the flowering of overtly feminist science fiction in the 1970s, extending into the 1980s, individual feminist works continued to appear, for example *Ammonite* by Nicola Griffith (1992). In 1991, the science fiction writers Pat Murphy and Karen Joy Fowler created the annual James Tiptree Jr. Award, in honor of Alice B. Sheldon, who published under the name

---

[4] This included a tendency away from emphasizing women's sexual liberation.

"James Tiptree Jr." (see Chapter 2). This has become an established and warmly regarded institution in the field of science fiction.

More generally, science fiction's iconography and tropes continue to provide enabling devices for authors to engage with a great range of philosophically important questions. These include ethical questions about war, economics, environmental degradation, personal morality, the human future more generally, the prospect of altering human capacities through technological means—and not least, the best principles for assessing all the above.

References

Asimov, I. (1981). *Asimov on science fiction*. Garden City, NY: Doubleday.

Bernal, J. D. (1970). *The world, the flesh, and the devil: An inquiry into the future of the three enemies of the rational soul*. London: Jonathan Cape (Orig. pub. 1929).

Blackford, R. (1985). The nest of the discorporator: Or, rereading *Stranger in a Strange Land*. In J. Blackford, R. Blackford, L. Sussex, & N. Talbot (Eds.), *Contrary modes: Proceedings of the world science fiction conference [sic] 1985* (pp. 61–78). Melbourne: Ebony Books in Association with the University of Newcastle.

Bogert, J. (1986). From Barsoom to Gifford: Sexual comedy in science fiction and fantasy. In D. Palumbo (Ed.), *Erotic universe: Sexuality and fantastic literature* (pp. 87–101). New York, Westport, CT, and London: Greenwood Press.

Huxley, A. (1958). *Brave new world revisited*. New York: Harper & Brothers.

Kass, L. R. (2001). Preventing a brave new world: Why we should ban human cloning now. *The New Republic* 21 May: 30–39.

Nozick, R. (1974). *Anarchy, state, and utopia*. New York: Basic Books.

5

Technophiles, Technophobes, and Renegades

Technology and the Uses of Power

Science fiction's emergence as a new genre or mode of fictional narrative was prompted by the revolutions in science, technology, and industrial production that took place in Europe during the seventeenth, eighteenth, and nineteenth centuries. These encouraged speculation about astonishing inventions and about future societies. Science fiction's main thematic concern is with the uses and consequences of advancing technoscience, but it does not follow that SF is always optimistic about science and technology. On the contrary, it often portrays lamentable uses of technoscientific power.

In the 2006 movie *X-Men: The Last Stand* (dir. Brett Ratner), Professor Charles Xavier (played by Patrick Stewart) makes a pertinent comment about the uses of power:

> When an individual acquires great power, the use or misuse of that power is everything. Will it be used for the greater good or will it be used for personal or destructive ends? Now this is a question we must all ask ourselves. Why? Because we are mutants.

In the X-Men mythos, mutants are people born with superhuman abilities.[1] Professor Xavier himself can project his thoughts, read the minds of others, and even control others' minds and thus bend their will to his. This is an

---

[1] More accurately, they are people with the genetic potential for superhuman abilities—the abilities usually manifest in childhood or adolescence.

© Springer International Publishing AG 2017
R. Blackford, *Science Fiction and the Moral Imagination*, Science and Fiction,
DOI 10.1007/978-3-319-61685-8_5

inherent personal power, but, within the ocean of narrative that constitutes the science fiction genre, great power takes many different forms. Again and again, SF writers warn against misuses of power that comes from science and technology.

The work of E.E. "Doc" Smith, stretching back to the Gernsback era of science fiction, exerted an incalculable and continuing influence. Smith's writing is melodramatic almost to the point of hysteria, but it contains much of interest. Smith created a stunning range of monsters, alien environments, and imaginary devices, and his villains were usually more interesting than his near-faultless heroes. Marc DuQuesne, the calm, cultured, supremely competent, and strangely honorable villain of the Skylark books, provided a template for many of the charismatic supervillains that followed in popular culture. Like some of those later villains, DuQuesne is both a worthy (often quite successful) antagonist and an invaluable ally for the heroes when the need arises to unite against worse evildoers. DuQuesne eventually saves the day in the final book of the Skylark series, *Skylark DuQuesne* (1966), written and published long after its predecessors. By the end, he is almost the hero—or at least the anti-hero—of the entire series.

Smith's overt theme is scientific responsibility, and particularly the responsible uses of powerful technology that can be employed for advanced weaponry. In *Galactic Patrol* (1937), from Smith's Lensman series, one of the mysterious and powerful Arisians explicitly sets out the moral stakes. He explains to Helmuth, the book's calculating and resourceful villain, that good and evil are relative, and that no culture is objectively evil. Helmuth's culture is, however, based on greed, hatred, corruption, violence, and fear; it does not recognize mercy or truth except for their utility; and it stands opposed to liberty of the person, thought, and action—which is said to be the basic goal of the opposing culture.

Even while denying that there is any any objective good and evil, the Arisian endorses a code of values. In consequence, Helmuth finds himself rebuffed in his mission to discover the workings of the Lens, a powerful device that the Arisians provided to the side of (relative) good in order to advantage it in a civilizational struggle against the piratical Boskonians. *Gray Lensman* (1939) contains a similar scene in which one of the Arisian superminds easily dispatches two Boskonian leaders. The Arisian then lectures the Boskonians' remaining leaders on their relative puniness (compared to Arisia) and on the illimitable power of the mind. All the Boskonians take from this encounter, of course, is the idea that thought can be a surpassingly powerful weapon.

As science fiction writers produce more complex characters, situations, and storylines, such elements take on lives of their own, and narratives become less schematic. They become, that is, more than straightforward demonstrations of the right and wrong uses of power. Even Doc Smith's rather Manichean space opera gives us villains—such as DuQuesne and Helmuth—who possess admirable qualities. Much science fiction appears to be searching for a standpoint from which we can acknowledge the subtleties of power and the choices it brings, while still achieving some clarity of moral vision.

Science Fiction's History with Technoscience

Questions about the moral status of technoscience reflect longstanding (sometimes opposed) tendencies in the Western culture that nurtured the science fiction genre in the first place (see generally Haynes 1994). In Medieval Europe, and within early Christian thought, there was a considerable tradition of suspicion toward anything resembling science, which was viewed as distracting or even idolatrous (Gaukroger 2006, 57–59, 151). This attitude was reflected in cautionary tales of magicians such as Faust. Cultural hostility to reason-based investigations of nature continued through the Renaissance, and it still found expression during the Age of Enlightenment in the eighteenth century. Jonathan Swift's satirical narrative of extraordinary sea voyages, *Gulliver's Travels*, first published in 1726, was in many ways a work of proto-SF. However, it expresses nothing but contempt for Enlightenment-era science. In his unintended voyage to Laputa, Lemuel Gulliver encounters scientists who are clearly shown as fools: they involve themselves in testing absurdly far-fetched hypotheses with no prospect of practical applications.

If anything, literary hostility to science and technology intensified during the Romantic period around the beginning of the nineteenth century, as with the depiction of Mary Shelley's anti-hero, Dr. Victor Frankenstein. As is well known, Frankenstein creates a grotesque living creature from the flesh of the dead. The subtitle of *Frankenstein*, "*or, the Modern Prometheus*," is somewhat ironical, since Prometheus was a benefactor toward mankind even though he overreached in challenging the gods. By contrast, Frankenstein does nothing to benefit mankind. He places it in danger with his powerful creation (and the threat, briefly, of creating an entire antagonistic species). He is Promethean only in his hubris. From its beginnings, then, science fiction showed a profound distrust of technology—and of nascent biotechnology in particular.

However, attitudes to science and technology changed during the second half of the nineteenth century. Technological progress came to be a popular value in industrialized nations, and this was reflected in the scientific romances of the time. There was considerable optimism about technological progress in

Jules Verne's tales of imaginary voyages, which postulated such wonderful devices as Captain Nemo's submarine in *Twenty Thousand Leagues under the Sea*. The period's techno-optimism culminated in futuristic utopias, such as Edward Bellamy's *Looking Backward: 2000–1887* and H.G. Wells's *A Modern Utopia* (1905). The more utopian component of Wells's output produced a backlash from other authors.

However, Verne and Wells understood the dangers, as well as the attractions, of powerful new technologies. Even Nemo's submarine is a powerful engine of destruction, used for purposes of vengeance as well as more admirable purposes such as deep-sea science. Nemo himself is not shown as heroic. Rather, he is better seen as an early—six decades earlier than Doc Smith's Marc DuQesne, for example—charismatic supervillain. Like many others who came after him, from DuQuesne to Alan Moore's Ozymandias and beyond, Nemo misuses his great power, yet can't be entirely condemned. He is partly admirable, thanks to the elements of sophistication and nobility in his character.

In *Metamorphoses of Science Fiction*, Suvin draws attention to the dangers that Wells reveals in science-based power:

> From the Time Traveller through Moreau and Griffin to Cavor, the prime character of his SF is the scientific-adventurer as searcher for the New, disregarding common sense and received opinion. (Suvin 1979, 210)

For Wells, as interpreted persuasively by Suvin, science is powerful in that it brings about the human future, but it is also dangerous, iconoclastic, and impossible to control.

Wells's Doctor Moreau of *The Island of Doctor Moreau* is a Frankensteinian figure—though there is also the suggestion of a kind of dark Galileo when we learn that he was hounded out of England after a public outcry against his research methods, receiving little support from his colleagues. Moreau attempts to create increasingly humanlike creatures from non-human animals; he produces a diversity of intelligent creatures, from bulls, to a puma, to a kind of hyena-pig hybrid. The methods he employs are horrifically painful for the animals involved, but as Moreau tells the narrator, Edward Prendick, he regards pain as of no consequence in the cosmic scheme of things. It does not deter him in his efforts to humanize animals and understand the plasticity of life. Moreau thus becomes monstrous in his quest for truth. Like Victor Frankenstein, he eventually comes to a bad end. He is killed by one of his own creations, a puma that escapes during his experiments on it. *The Island of Doctor Moreau* is an uncomfortable story, notable for its close observation and realism as it describes the hideous situation and disturbing events on Moreau's isolated island.

Among the scientists in Wells's early scientific romances, Moreau is matched as a figure of evil only by Griffin, the central character of *The Invisible Man* (1897). Griffin has succeeded in turning himself permanently invisible, expecting this to give him great advantages in life, but he finds to his exasperation that it brings more in the way of disadvantages, cutting him off from most ordinary interaction with others (he is forced to walk about either carefully masked and muffled in some way or else totally naked). Griffin is a big, physically powerful, violent man with a psychopathic personality. As he discovers, invisibility does give him some advantages—certainly in acts of theft, but especially in inflicting violence. He becomes a social menace who leaves behind a trail of dead or battered victims. This includes one especially brutal murder in which he uses an iron rod.

The Invisible Man asks how human beings would—and should—employ great power granted to them via technology. Griffin uses his power for selfish ends and employs notably callous methods, seeming to regard all others around him as merely impediments to his desires: as fools and blunderers who get in his way, or at best as individuals whom he might be able to manipulate. His invisibility helps him commit crimes and, for a time, get away with them. In that sense, he possesses an equivalent of the Ring of Gyges discussed by Plato in his dialogue the *Republic* (originally composed c. 380 BCE). However, his inability to turn visible and deal with others in a more normal way also disables him. It places him in a particularly difficult position, since the *only* advantage that he obtains is an enhanced ability to commit crimes of stealth or violence.

Griffin's invisibility isolates more than it empowers him. Yet, his personality is shown as narcissistic, anti-social, and ruthless even before he succeeds in turning invisible. Unlike the (initially) gentle monster created by Franken-stein, Griffin was always monstrous. This prompts the question of how someone more benign might have reacted to his predicament. Note, however, that even the eccentric and likeable Mr. Cavor, from Wells's *The First Men in the Moon*, appears somewhat Faustian and pays dearly for his technoscientific exploits. Cavor ends up stranded on the Moon—and is possibly killed by the Selenites—as a result of discovering the anti-gravity substance Cavorite. He is almost as obsessed as Moreau with the quest for scientific knowledge, though he shows nothing of Moreau's inhuman willingness to inflict pain in order to achieve his ends.

Overall, the scientists and inventors in Wells's early scientific romances don't come out of things in good shape. Even the redoubtable Time Traveller in *The Time Machine* gains only trouble, heartbreak, and despair from his stunning invention. The narrator of *The War of the Worlds* does much better, but he is a scientifically informed speculative philosopher, rather than a scientist, and unlike the Time Traveller, Moreau, and Cavor he does not initiate the terrifying course of events.

The more optimistic side of Wells's vision was reflected in the magazines of the Gernsback and Campbell eras. In science fiction narratives of the Golden Age from the late 1930s to the end of the 1940s, scientists, engineers, astronauts, and other men (less often women) who are skilled in technoscience become heroes when they improvise creatively and overcome technical limits. As I elaborated in Chapter 2, Isaac Asimov's fiction epitomizes the Golden Age's emphasis on logical problem solving and its optimism about the uses of science and technology. By way of contrast to this approach in the SF magazines, the dystopian possibilities for science fiction were demonstrated by two novels written around this time for general audiences rather than the specialized SF market.

Aldous Huxley's *Brave New World* and George Orwell's *Nineteen Eighty-Four* depict, respectively, a dystopian future of biologically engineered harmony that is portrayed as shallow, bland, and dehumanized, and a dystopian future of unprecedented, utterly brutal totalitarian control. *Brave New World* has become synonymous with the nightmarish uses of advanced reproductive technology in order to control the populace. Huxley shows how a biologically-sustained class system depends, in part, on the "Bokanovskification" technique used to create batches of identical babies (though not to replicate existing humans as usually contemplated in discussions of reproductive cloning). Science and technology are deeply complicit in the regimes of control imagined by Huxley and Orwell.

Early science fiction movies tended to be Frankensteinian or dystopian, as with Fritz Lang's *Metropolis*. This offers a disturbing vision of a mechanized and dehumanizing future city, complete with a mad scientist (Dr. Rotwang) and a lifelike robot (commonly known as "the false Maria") as its formidable villains. However, some early movies were more optimistic, most notably *Things to Come*, based on works by Wells.

Like Gaul as it was once described by Julius Caesar, *Things to Come* is in three parts. The first depicts a near-future war that continues for decades. Second, we see the success of the Airmen in taking over, after the world has become barbaric and decadent, ruled by local warlords. The third part imagines enormous possibilities for science. Here, we are shown the utopian rule of technocrats, descended from the Airmen, in the twenty-first century. As the movie concludes, the technocrats succeed in firing their space gun to launch a flight around the Moon, despite the opposition of an artist-cum-demagogue and his rabble of followers.

Seen eighty years later, *Things to Come* appears like one-eyed propaganda for an overweening form of science. This was, of course, how it was viewed in its day by hardline anti-Wellsians such as C.S. Lewis.

Darker themes emerged in science fiction after World War II. Post-war SF began to express fears of decay or destruction, produced by over-reliance on technology and/or the use of immensely powerful weapons. This was influenced by broader cultural trends: the bleak tendencies in 1950s and 1960s science fiction were part of a widespread distrust of technoscience after the nuclear destruction of Hiroshima and Nagasaki, and the technologically facilitated atrocities of the Nazi death camps.

A common post-war theme across all media was that human beings are not yet ready to control the advanced technologies that could destroy us and our world. *Forbidden Planet* is a good example, with its "monsters from the id" which (so it transpires) destroyed an entire alien civilization, the Krell. The tragedy is repeated, in part, when Dr. Morbius, who is marooned on the forbidden planet with his beautiful daughter, Altaira, gains control of the same unstoppable power. During the 1960s, British and American authors of the New Wave era often expressed visions of a doomed society. This is exemplified in such works as *The Wind from Nowhere* (1961), *The Drowned World* (1962), and *Crash* (1973), all by J.G. Ballard, a dominant figure within the British New Wave.

A strain of pessimism has continued through literary, cinematic, and televisual science fiction. However, *Star Trek* maintained an opposed tradition on television during the late 1960s. It featured the voyages of a giant starship, sent out to explore for new worlds and civilizations. However clumsily, *Star Trek* suggested the possibility of a future space-faring society without racism, sexism, poverty, or gross inequality. Optimistic visions of technoscience also appeared in the prose medium, especially in novels and stories in the tradition of hard science fiction. Hard SF authors such as Gregory Benford, Greg Bear, and David Brin elaborated and deepened the Campbell-era emphasis on the problems encountered by working scientists and other skilled professionals.

Through the late decades of the twentieth century—and continuing into the twenty-first—the technophobic side of SF focused increasingly on dangers from computers, artificial intelligence, and biotechnology. However, the 1980s cyberpunk movement handled all this with a degree of ambivalence or detachment. Cyberpunk and cyberpunk-influenced narratives showed technology interfacing directly with the brain or otherwise penetrating the body. Old notions of the body in space and time were replaced by images of the body's transformation via technological systems such as prostheses and advanced computers. Cyberpunk writers showed emerging technologies as potentially damaging to societies and individuals—but at the same time, as inevitable, adaptive, and in some ways attractive. Near-future cyberpunk novels such as William Gibson's iconic *Neuromancer* are not totally dystopian. They invite neither total condemnation nor total approval of the societies that they portray.

The lesson, perhaps, is that technology is not entirely good or entirely bad. It is, however, inevitable, increasingly ubiquitous, and always likely to find unexpected uses. We now swim in a sea of technological gadgets and systems, without which individual human beings would be very different—and our civilization could never survive. Cyberpunk reveals this, without necessarily moralizing about it.

A similar lesson applies to science fiction more generally. Part of SF's appeal lies in its ability to show advanced technology as dangerous and potentially destructive, while simultaneously revealing its allure and giving it some accommodation within an implied scheme of values. Thus, the critic J.P. Telotte refers to what he calls "the sort of double vision" typifying science fiction films. He elaborates this as "a tendency to accept but also draw back from the alluring technological imagery that empowers the science fiction film and that finds specific embellishment in every robot" (Telotte 1995, 126). One of Telotte's examples is the ending of *Forbidden Planet*, where we do—literally—draw away from the planet and watch as it is destroyed. But some remnants or reminders of the forbidden planet survive, notably the powerful robot that Morbius created by means of Krell technology. Along with the robot, Robby, goes the beautiful, wild Altaira. She will now adapt to—and at the same time be accommodated by—the space-faring human civilization of the future.

Contemporary science fiction shows a wide range of attitudes to technoscience. Some authors evidently see it as a blessing, and others as a curse. Optimists imagine the development of a more mature, less warlike, humanity, the wise integration of science into society, and expansion into space. Pessimists emphasize the dangers of technoscience, often illustrated by narratives set in unpleasant future societies—and even more often dramatized by the rampages of new kinds of monsters. As we think about these, however, it is worth keeping in mind Telotte's point about a double vision.

In Hollywood's SF blockbusters, even the most dangerous products of technology, such as the killer cyborgs portrayed in *The Terminator* and its sequels, are not merely demonic. Eric G. Wilson (2006) points out that movies such as *The Terminator* do not *merely* allow us to live out our fears of machines, while providing some reassurance when they are defeated. These movies enable us to identify secretly with nonhuman intelligences that are displayed as untroubled in their rampages. The Terminator acts without hesitation. It experiences no gap—no shadow of anxious choice—between desire and action (see Wilson 2006, 155–56). This is monstrous, but for creatures like us, plagued by ambiguity, self-doubt, and hesitation, it also seems enviable. The cyborg assassin played by Arnold Schwarzenegger is, in its way, *cool*—as are many of science fiction's monsters, anti-heroes, and supervillains.

Science Fiction Meets Bioscience

From its beginnings, science fiction has postulated advances in biology, and has depicted the work of biological scientists, the biology of alien beings and alien worlds, and mutational changes on Earth. Its specialized concerns have included artificial organic life, medical advances, new diseases, genetic engineering, and the creation of human or animal clones. Despite this variety of topics, physical or moral monstrosity typically results from human meddling with the stuff of life. Discussing science fiction in the cinema, Roslynn D. Haynes describes the genre's harsh representation of biologists:

> Less cataclysmic and hence more contained than the nuclear threat is that posed by the evil biologist, direct heir of Frankenstein, who embarks on sacrilegious attempts to change existing life forms, either by destroying them or by shrinking, enlarging or transmuting them. This scenario has changed very little over the years from "radium rays" to *Jurassic Park*, for in the moral world of cinema such tampering with the sanctity of life and species uniqueness is never condoned and retribution invariably follows. (Haynes 2000, 42)

Frankenstein's monster is especially disturbing and creepy precisely because he is not a robot as that is understood today. The monster is not an artificial creature of metal, but one cobbled together from once-living flesh. More generally, biotechnology seems to be especially troubling to the technophobic imagination. All in all, science fiction has usually taken a negative approach to imagined biotechnologies. This applies especially to works that command very large audiences, such as Hollywood movies.

Gregory E. Pence, an American bioethicist, has argued that science fiction portrays human cloning, in particular, in a negative light, contributing to a widespread fear of cloning technology (1998, 39–43). Some of Pence's specific criticisms and examples could be challenged, but on the whole his argument is persuasive. Science fiction authors do seem to regard the replication of humans or animals, through cloning techniques, with a special repugnance. To be sure, there are exceptions. For example, in "Houston, Houston, Do You Read?" (1976), by James Tiptree Jr. (Alice B. Sheldon), a NASA spaceship is thrown into the future, where its male crew members encounter a feminist utopia of women who are all clones. In this society's past, a plague prevented male births and forced a drastic technological response. This society is portrayed sympathetically, whereas the men who are confronted by it come off badly.

More typically, however, cloning is presented as an abuse of technology. An extreme example from the 1990s is Michael Marshall Smith's *Spares* (1996), which depicts a surrealistic and horrifying future society in which human cloning is used as a source of spare body parts. In this gruesome dystopia, biotechnology is used to provide twins each time a new child is conceived and born: one child grows up normally to take his or her place in society, while the other is confined to a farm for "spares." The spares are subjected to the most inhumane conditions, treated worse than farm animals. If a spare limb or organ is required when the socialized child meets with illness or accident, the appropriate part is simply removed from the spare—without even the use of anaesthetic. A similar idea appears in the 2005 movie *The Island* (dir. Michael Bay).

In the Jurassic Park novels and movies, cloning is used to recreate ancient reptiles from fossilized, and incomplete, DNA. This goes terribly wrong, and the implication seems to be that we'd better not "play God" by tampering with nature. There is also a morality play element in this series, often highlighting the characters' attitudes to technology. Many characters are killed swiftly—they are pretty much treated as dino fodder—but elaborate, and often humiliating, deaths are given to the characters who appear most venal or blinded by pride. Perhaps the most humiliating death of all is given to the lawyer, Donald Gennaro, in the first movie of the series, *Jurassic Park*. Interestingly, this scene does not appear in the 1990 Michael Crichton novel, also called *Jurassic Park*, on which the movie was based. Presumably it was inserted to pander to popular resentment of lawyers.

The more sympathetic characters in the Jurassic Park movies are shown as having moral weaknesses, but they are punished by their terrifying encounters with the rampaging dinosaurs and ultimately redeemed.

The genetic technology used to reconstruct dinosaurs from fossilized DNA is fairly consistently portrayed as evil. The Jurassic Park movies invite us to believe that the exercise of creating the dinosaurs from ancient genetic material is an act of hubris. But there is an ambiguity here, a certain instability of attitude, because the dinosaurs themselves are not merely dangerous and terrifying. Some are relatively harmless, and they are shown as fun and exciting, or even as majestic or sublime. We, as moviegoers, are much like the audience of the theme park in *Jurassic World* (dir. Colin Trevorrow, 2015): we expect to be impressed and awed by the dinosaurs, not merely scared by them. It follows that the dinosaurs are not portrayed straightforwardly as monsters. To a large extent, they are instruments of fate, inflicting rewards and punishments. In a sense, the *real* monsters of *Jurassic Park* and its sequels are the human beings who exploit genetic technology in ways that appear greedy, vain, and irresponsible.

These elements are handled with a certain knowingness in *Jurassic World*—the most recent installment in the series as I write these words, although a further sequel is scheduled for release in 2018. *Jurassic World* introduces a frightening new dinosaur, *Indominus rex*, which is *not* an attempt at recreating a beast from the Mesozoic Era. It has been genetically engineered as a theme park attraction that will be more impressive than the likes of *Tyrannosaurus rex*. *Indominus rex* is an almost demonic creature, notable for killing other dinosaurs for sport. With its deliberate "improvements," it is made to seem even more unnatural than the other artificially created dinosaurs. To rub in just how malevolent this creature is, its enhanced abilities include extraordinary stealth and cunning, as well as the dialed-up cruelty that was requested in its specifications.

Of course, the portrayal of biotech as Frankenscience is not confined to science fiction. In the real world, there is much resistance to innovations in biotechnology. Science fiction writers are unusually sophisticated in their knowledge of science, so we might ask what role they should play in public debate over bioethics and biopolitics. Regrettably, SF works aimed at a wide audience, and particularly Hollywood movies, are likely to do no more than reflect and encourage bioluddite sentiment.

Frankensteins, Promethean Figures, and Galilean Scientists

In his 1996 article "The Science Fiction of the House of Saul: From Frankenstein's Monster to Lazarus Long," Barry Crawford concentrates on what he sees as the figure of the rogue male: a dissenting individual of great ability who is isolated from society, unwilling to obey its customary dictates, and often shown seeking forbidden kinds of knowledge and power, perhaps including some form of immortality. On Crawford's account, such figures are depicted in holy books (as in the story of Saul) and literature (as in *Frankenstein*) as transgressors against the social order. Their traditional narrative role is to be defeated by others with more domesticated and orderly desires. According to Crawford, however, some science fiction writers—he focuses on A.E. van Vogt and Robert A. Heinlein—openly take the side of the rogue male. What should we make of this?

Some of it strikes me as misguided, but Crawford has a point. His primary examples are Heinlein's depiction of the immortal and adventurous Lazarus Long, in *Methuselah's Children* (1958)[2] and *Time Enough for Love* (1973), and certain characters in van Vogt's *Slan* (1940) and *The Weapon Makers* (1943; revised version 1952). Lazarus Long, who also appears in Heinlein's *The Number of the Beast* (1980), *The Cat Who Walks Through Walls* (1985), and *To Sail Beyond the Sunset* (1987), is undoubtedly a character who defies many conventions and transgresses Christian *mores*. He can be seen as one of the secular Elect figures (see Chapter 4) who appear frequently in Heinlein's work.

Slan is van Vogt's classic story of superior beings walking among ordinary humans, with resulting conflict and atrocities. The slans possess more-than-human intellect, strength, stamina, and ability to recover from injury. The original slans and their descendants also possess telepathic powers, based in the antenna-like tendrils growing from their heads. These tendrils conspicuously mark the slans apart as a separate breed. However, the novel's main protagonist, the young slan Jommy Cross, soon discovers that there are slans who do not possess tendrils or telepathy. These have created their own powerful and militarized society. Though disadvantaged in not being telepathic, the "tendrilless slans" have a great advantage compared to Jommy and his kind in being able to pass as ordinary humans. The slans are depicted as persecuted and heroic, and *Slan* can be read at one level as allegorizing racial and ethnic persecutions, such as those experienced at the time by the Jews in Nazi Germany.

In van Vogt's *The Weapon Makers*, we are invited to side with the isolated immortal Captain Robert Hedrock. Employing a mix of cunningly laid plans, decisive action at moments of crisis, and super-scientific weapons and other devices, Hedrock pits himself against the solar system's great political powers. His ultimate goals are to kick-start interstellar colonization and to discover (and make generally available) the secret of his own immortality. Accordingly, his narrative arc glorifies an ethic of human destiny.

---

[2] *Methuselah's Children* first appeared in serial form in 1941. It was expanded for book publication in 1958.

Crawford is undoubtedly correct that science fiction is well stocked with renegades who transgress the established order, often using or misusing advanced technology to do so. Most of them come to a bad end. Victor Frankenstein is clearly such a figure, just as Crawford says, and so are many of Wells's scientists and inventors, including Moreau, Griffin, and perhaps even the insouciant Cavor. Most of these characters are male, and science fiction narratives take varied attitudes to them—often portraying them as dangerous, even if they fundamentally mean well. Crawford is also correct that genre science fiction has often treated similar figures as heroic. He is critical of much 1980s feminist SF criticism as an attempt to domesticate the rogue male figure—thus taking the side of the more genuinely oppressive men who uphold the established order:

> In seeking to denigrate or suppress the rogue male who threatens the existing (patriarchal) order, feminists almost become allies of patriarchy and fight with literary critical tools fashioned in an essentially Davidic world of moral sophistry. (Crawford 1996, 155)

This is all fascinating, but it's best to tread carefully in this territory. There is no deep reason why the rebellious figures that Crawford describes must always be male—or why, in science fiction narratives, they must even be human characters. For most of its history, science has been viewed as largely (if not entirely) a male preserve, but women are increasingly entering the sciences, and some doubtless possess the sort of personality that Crawford describes: impatient with authority; fiercely committed to truth even to a point where the quest risks self-destruction; and focused, perhaps, on culturally taboo kinds of knowledge.

In our actual, non-fictional world, bold-thinking scientists have frequently encountered difficulties when they've challenged the established order or otherwise violated taboos. Galileo Galilei, who was punished by the Inquisition in 1633 for his defense of heliocentrism, is the archetype of this kind of scientist. In her 2015 book *Galileo's Middle Finger*, Alice Dreger discusses the ordeals of a number of contemporary scientists who have been persecuted in one way or another for heretical or taboo ideas. In each case, these individuals were not reckless troublemakers getting their thrills by watching the world burn. They were careful, socially aware researchers whose substantive viewpoints were rejected by opponents for reasons having as much to do with ideological preconceptions as with empirical evidence. In some of these cases, nonetheless, their willingness to pursue transgressive approaches appears to have been accompanied by abrasive personalities.

Since Galileo is the obvious archetype for this sort of figure, Dreger has introduced the term "Galilean personality." In *Galileo's Middle Finger*, this first appears in a passage where Dreger is describing the anthropologist Napoleon Chagnon:

> But exacerbating tensions was the fact that Chagnon had the classic Galilean personality, complete with political tone dumbness—that inability (or constitutional unwillingness) to sing in tune. Indeed, descriptions of Chagnon provided to me by both his friends and enemies sounded eerily reminiscent of Galileo: a risk-taker, a loyal friend, a scientist obsessed with quantitative description, a brazen challenger to orthodoxy. (Dreger 2015, 141)

As Dreger describes them, such Galilean types, or Galilean personalities, provoke trouble—and tend to get into trouble—not because they seek it but because they insist on the truth as they understand it. They end up paying a price in their own suffering. The general theme of *Galileo's Middle Finger* is an insistence that we ought to value such people and not shirk from expressing the Galilean sides of our own personalities. Dreger's examples are male, but she has no hesitation in viewing herself as another of these Galilean types.

Again, science fiction abounds with such figures, but there is a thin line between a Galilean personality—or, even more grandiosely, a Promethean one—and a Frankensteinian personality. Moreover, many different kinds of characters can show elements of these related character types.

Felix Hoenikker, the scientist whose invention, ice-nine, ultimately destroys the world as we know it in Kurt Vonnegut's *Cat's Cradle*, is clearly presented as an updated Frankenstein. He epitomizes much of what Vonnegut evidently saw as worst in modern Western society, in particular the search for knowledge without considering its human effects. Another such Frankensteinian figure for the nuclear age is the eponymous Dr. Strangelove (played by Peter Sellers) in *Dr. Strangelove or: How I Learned to Stop Worrying and Love the Bomb* (dir. Stanley Kubrick, 1964). As an example of intertextual influence and reference, Dr. Strangelove's creative design owes something to Dr. Rotwang's in *Metropolis* (Frayling 2006, 26).

For his part, Vonnegut offers us an alternative in *Cat's Cradle*: the holy fool Bokonon. Like Winston Niles Rumfoord in *The Sirens of Titan*, Bokonon is the creator of a bogus religion designed to improve the lot of humanity. He is an imposter, but he freely admits this. Bokononism does not exist to provide any genuine explanation of the universe, but to make the human situation more bearable for oppressed souls. Readers of *Cat's Cradle* are not asked to become programmatic Bokononists, but merely to share Bokonon's compassion rather than Hoenikker's search for scientific knowledge at all costs.

Like Bokonon, the vengeance-consumed Gully Foyle from Alfred Bester's *The Stars My Destination* is not a scientist or an inventor, but by the end he does become a Promethean figure of sorts. He makes the powerful substance PyrE available to humanity, leaving the consequences undetermined. His action might lead to a greatly empowered humanity or to humankind's destruction, so it's unclear whether he is a Promethean benefactor or the antithetical opposite.

Winston Smith, from Orwell's *Nineteen Eighty-Four*, is likewise not a scientist; however, he shows something of the Galilean personality as Dreger describes it. While his lover, Julia, seems most revolted by the sexual puritanism imposed by the Inner Party, and is only secondarily concerned by the neverending propaganda pumped out by the regime of Oceania, Winston has the opposite priorities. He cannot ignore the knowledge—available from his own memory and from his job as a propagandist in the Ministry of Truth—that the regime is systematically falsifying and rewriting history. Only later in the story does his growing love for Julia become overwhelmingly important to him. Their rebellion against the established order never amounts to much, and it is easily crushed by the Party apparatus, but at least Winston represents a type of person who will not lightly let the truth go, no matter how much it seems in his interests to conform.

Isaac Asimov's *The End of Eternity* involves at least two figures with distinctly Galilean, if not Promethean, tendencies, and in this case one of them is female. The main character, Andrew Harlan, is a Technician employed by the all-powerful, somewhat monastic, organization Eternity, which I discussed in another context in Chapter 4. Recall that Eternity holds the secret of time travel, and it uses this technology to make continual adjustments to the course of human history over thousands of centuries. The appearance of monasticism comes partly from the way Eternity is structured, but also from the fact that it is an all-male organization. It is staffed by individuals recruited from various points in history. Women are not chosen because the effect of removing individuals from time is greater if they're female than male.

Harlan is something of a lone wolf within the organization. He becomes even more self-absorbed, paranoid, and dangerous when he falls in love with Noÿs Lambent, a woman from (as it appears) the 482nd century. In his efforts to protect her and to save his love affair—which goes against Eternity's rules—he devises a plot to bring down Eternity itself. Only after he has been dissuaded from this course of action does he discover that Lambent is not what she seems. Instead, she is a woman from further in the future than any period that Eternity has dealt with. Her mission is to destroy Eternity, and she eventually succeeds. In the process, she makes her own massive change to history by introducing nuclear weapons into the twentieth century: she is prepared to risk the devastation of Earth itself for the benefits from hastening exploration of space. In acting so dramatically to reshape human history, Lambent becomes the ultimate techno-transgressor, certainly eclipsing her malcontent lover who is brilliant but often misreads situations.

For a purer example of the Galilean scientist, we might consider Shevek in Le Guin's *The Dispossessed*. Recall (from Chapter 4) that *The Dispossessed* is set on two worlds that orbit the star Tau Ceti. Urras somewhat resembles twentieth-century Earth, with a variety of nations operating under different political and economic systems. Of these, the dominant nation is A-Io, something of a caricature of the USA, with advanced science and technology, great material plenty, and a seemingly utopian lifestyle for its propertied classes—all accompanied by injustice, since it is built on the poverty of A-Io's proletariat. This type of dystopian society resembles that described in *When the Sleeper Wakes* and that shown on the cinema screen in *Metropolis*. The women of A-Io are excluded from higher education and from positions of leadership, though women from the propertied classes are pampered and enjoy considerable personal freedom. Within the family setting, they are sometimes treated by their husbands as friends and equals, but the society as a whole reeks of sexism.

On Anarres, rebellious colonists have established their own civilization along anarchist lines. By contrast with the abundant natural resources of Urras, Anarres is a desert world. Human life is made possible only by continual struggle and effort, and through trade—albeit minimal and much resented—with Urras. Anarres was founded as an anarchist utopia, with various coordination mechanisms but no state apparatus or formal laws; however, it has seriously degenerated. Its morality of community contribution and mutual aid has largely given way to a tyranny of public opinion and cunning, small-minded bureaucrats. It has become a conformist society whose members continually police each other for moral lapses, especially the dreaded "egoizing" (anything that can be construed as showing off or seeking attention).

In this setting, Shevek, a brilliant and non-conforming physicist investigating the nature of time, finds his work continually frustrated by his supposed mentor, Sabul, and by the scientific bureaucracy that enables him. Shevek is repeatedly told that his most fundamental research has no practical application, and hence offers no social benefit. As we learn, however, its practical implications are actually staggering. Shevek's theory provides the basis for instantaneous communication even across interstellar distances, circumventing the hard physical limit on faster-than-light travel and transmission of information.

Disillusioned, Shevek leaves Anarres for Urras, enabling Le Guin to employ him as an alien observer of A-Io society in particular. Thus he sees its splendor and its undoubted power, but he also discovers its darker side. Although Anarres has degenerated and taken on its own dystopian tendencies, the journey to Urras at least clarifies Shevek's loyalty to what Anarres attempted. During his time on Urras, Shevek repeatedly defends Anarres in arguments, even though he knows by this point that Anarres is, in reality, something less attractive than his idealized picture of it.

As *The Dispossessed* concludes, Shevek does not offer his theoretical findings to Anarres or to any of the societies of Urras. Instead, he provides them to representatives of other planets, and thus lays the scientific foundation for divergent branches of humanity scattered across interstellar space to communicate and coordinate in real time. This will alter the entire direction of humanity's future. We don't see the consequences in *The Dispossessed*, although the resulting communications device, the ansible, appears in other work by Le Guin. In any event, Shevek has become a Promethean benefactor to humankind.

At the same time, just as Shevek and his immediate circle of supporters hoped, his bold gesture in traveling to Urras encourages dissenters from the stifling environment of conformity on Anarres. It pushes them to challenge the existing bureaucracy. This confirms that the original ideals of Anarres were of value and that Anarresti society retains the resources to renew itself.

Gregory Benford: Nigel Walmsley and Other Renegades

Gregory Benford's fiction provides a useful case study, since it often focuses on the thoughts and actions of sympathetically portrayed Galilean scientists: characters who fairly closely match the Galilean type. These individuals are driven by their desire for truth, and they show a renegade streak. The prime example is Nigel Walmsley, whose personality and choices dominate Benford's Galactic Center Saga. The novels in this series are, in order, *In the Ocean of Night* (1977); *Across the Sea of Suns* (1984); *Great Sky River* (1987); *Tides of Light* (1989); *Furious Gulf* (1994); and *Sailing Bright Eternity* (1995). Walmsley is the protagonist of the first two and an important presence in the sixth and last.

In The Ocean of Night deals with the gradual realization, in the late twentieth and early twenty-first centuries, that humanity is not alone in the universe. We are, rather, one of many forms of intelligence, some biological and others computational, and a vast struggle for survival and power is going on around us. In this first volume of the Galactic Center Saga, the nature of the struggle is only hinted at or, at most, sketched. We do see, however, that it involves a long-running war between mechanical and biological beings.

Walmsley is introduced as a US-based British scientist with a mind of his own: he defies orders and authority on several crucial occasions, beginning in 1999, when he is sent on a mission to destroy and deflect Icarus, an asteroid on a collision course with Earth. He discovers that Icarus is not a natural object but the relic of an alien space vehicle. Though the risk of destruction to Earth is increased the closer Icarus approaches, Walmsley delays, believing that this will give an opportunity to explore the interior of Icarus, looking for advanced technology that might solve some of humanity's problems (some alien technology is, in fact, discovered, though nothing that causes a scientific breakthrough).

In 2014 (and through to 2019 when the narrative closes), an older Walmsley continues to act as something of a rogue figure, taking his own steps to contact a new alien intruder (rather than implementing the approved work of his team), refusing to (attempt to) destroy it when ordered, releasing data on an alien artifact discovered on the Moon, and generally trusting his own instincts and deductions above those of his colleagues and superiors. Walmsley comes up against skilled administrators, but their thinking is safe and limited. Despite getting into difficulties, he is able to survive because he is at least partly vindicated on each occasion. By the end of *In the Ocean of Night*, he has obtained an almost mystical understanding of the universe and what must be done to investigate it.

Walmsley is difficult, self-confident, obsessed by the search for truth, and willing to buck the system and offend powerful individuals when needed to pursue that search. He shows a touch of ruthlessness or arrogance, in that he takes risks that could affect others as well as himself—though they also offer possible benefits for everybody. Like Galileo, he meets opposition from religious quarters: in this case the intellectually subtle and politically powerful New Sons. As he sees things, organizations can contain many individually fine people but organizations also have drives of their own that can get in the way of the truth. When another character reminds him of the need to compromise, Walmsley's response is a grimace and a slamming door as he leaves the room.

Across the Sea of Suns takes place about four decades after the conclusion of *In the Ocean of Night*. Shortly after the events of the earlier book, Earth-based governments send out an interstellar ramscoop ship, the *Lancer*, to investigate the presence of alien life in nearby star systems, beginning with a small red star from which non-random electromagnetic signals have been detected. Aboard the *Lancer* are Nigel Walmsley, who is elderly and physically weaker by now, and his lovers. As in the previous book, Walmsley butts heads with other characters, sometimes making mistakes (more in his personal life than his scientific quest, since he tends to sacrifice the former to the latter). Meanwhile, he single-mindedly investigates his hunches, defying authority and consensus when needed. Overall, his analyses are vindicated as he develops the picture of a great interstellar struggle between organic and mechanical forms of life. He also tries to understand the behavior of human beings—and particularly what he sees as the distinct emotional tendencies of men and women—in the ship's very high-tech and socially exploratory society.

Great Sky River, *Tides of Light*, and *Furious Gulf* take place in the very distant future long after the fall of human civilization in its various manifestations. Only scattered remnants of humanity are left in the universe. The dominant figure in these novels is Killeen Bishop, a brilliant intuitive leader who is another somewhat Galilean figure. He is, as the critic George Slusser puts it, "a future human who possesses the same trait as Nigel." Slusser continues, a few sentences later, "Killeen is a future human being who is physically and mentally 'enhanced' yet by the force of things is compelled to revert to primitive tribal conditions." He shares Nigel's "intuitive vision of the vastness of the cosmic mechanism" (Slusser 2014, 85).

Killeen and his group are caught up in wars against the "mechs" and other parties. In *Tides of Light*, they encounter a race of gigantic—somewhat insectoid, somewhat plantlike—cyborgs, who are divided among themselves as to how to treat humans: whether to nurture them as important to their ongoing struggle against the mechs, or exterminate them as dangerous pests. Much of the story is told from the viewpoint of Quath, one of the cyborgs, who turns out to have a probing, anxious mind of her own as she rises to leadership within the hierarchy of her people.

In *Furious Gulf*, Killeen leads his followers to the Galactic Center, but the focus switches to his son, Toby, who has an encounter with the still-living Nigel Walmsley. *Sailing Bright Eternity* allows us to catch up with Walmsley's doings after he left the scene at the end of *Across the Sea of Suns*. Nigel talks with Toby Bishop and offers his own explanation of events. We learn that Nigel has been as willing as ever to take risks and to question received wisdom. We also meet our old friends Killeen and Quath.

Throughout the Galactic Center Saga, the characters who advance truth and human welfare—and the welfare of biological beings in general—show curiosity, creativity, and willingness to rebel when it is needed to uncover the truth. This applies to the alien cyborg Quath as much as to the human characters. Benford suggests that this type of personality is inherent in human nature. It is not that we all possess it, but it emerges in some individuals in all societies. Indeed, Quath discovers that her people carry strands of DNA from human beings, contributing to their curiosity and creativity.

While the Galilean scientists, and similar figures, in science fiction are usually male, the cyborg Quath is female. Furthermore, Benford's female scientists often have similar character traits. We see this with Julia Barth, for example. She is the astronaut and biologist who figures as the main character in *The Martian Race* (1999)—not a book about a race of Martians, but the story of a technological and astronautical contest to get humans to Mars. It leads, in its sequel, *The Sunborn* (2005), to a rivalry between two memorable female characters: Julia, again, and the younger Shanna Axelrod. Both women are willing to make consequential and controversial decisions. Despite their mutual prickliness and resentment, they recognize each other as the sort of people who take action, draw strong conclusions, and frequently get into trouble. One implication running through Benford's fiction is that we *need* such people for the future of humanity.

Cixin Liu: Renegades with Responsibilities

Remembrance of Earth's Past, by Cixin Liu (Liu Cixin), is a trilogy exploring the themes of death, love, and responsibility. Liu has created many unforgettable characters who face factually and morally complex decisions. Their personalities are themselves complex, but they are drawn with some bold, vivid strokes. Many are, to an extent, rogues or renegades—refusing to act as is socially expected of them—but they are placed in such extreme situations, often with the fate of humankind and alien worlds at stake, that even their most drastic choices are understandable.

In the first novel of the trilogy, *The Three-Body Problem* (2006; English translation 2014), we meet the embittered astrophysicist Ye Menjie. Her bitterness is grounded in experiences of persecution by the Chinese authorities, including witnessing her father's death at the hands of Red Guards in 1967: early in the novel, he is beaten to death by teenage fanatics for teaching supposedly reactionary Western science, such as Einsteinian relativity theory. Years later, in 1979, Ye Menjie takes the fatal step of calling down an alien invasion on Earth, foolishly hoping that contact with an alien civilization will reform humanity. Though she is a tragic and sympathetic figure in some ways, she is more Frankenstein than Prometheus. She is no benefactor; instead, she places our world in peril.

By contrast, Luo Ji, the main character of *The Dark Forest* (2008; English translation 2015), acts for humanity's benefit. Enigmatic, hedonistic, but possessed of deep insight, he is tasked with saving Earth from the impending invasion while devising a plan that cannot be detected by the alien Trisolarans' surveillance technology. In doing so, he is prepared to put both civilizations at risk, but from Earth's viewpoint there is no other choice, as the Trisolarans, with their incomparably superior technology, intend to exterminate humanity like so many bugs.

The final book in the trilogy, *Death's End* (2010; English translation 2016) is centered on a young scientist, Cheng Xin, who comes to inherit the task of Luo Ji in guarding humanity's fate. She takes on this crippling responsibility partly from being in the right (or wrong place) at the right time, but largely through her intellectual brilliance and her highly attractive personality. Cheng Xin has an obviously kind and loving nature that wins trust and attracts responsibility whether or not she seeks it. But she acts out of love, even when this seems naive or irrational within the horrifically bleak universe that was revealed to us in *The Dark Forest*. On two occasions, she makes what seems the "wrong" decision for preservation of humanity as a species, and yet the final effect of *Death's End* is to question whether these were wrong decisions after all.

On the first occasion, others make the "right" choice when she fails to: they broadcast to the larger universe the location in space of Earth's Trisolaran enemies, leading to the destruction of the Trisolaran home world. On the second occasion, she averts a catastrophic war within Earth's solar system, while also taking away Earth's possible hope of developing lightspeed space travel. On this occasion, it is simply not clear whether she acted wisely. The alternative choice might or might not have saved Earth from its eventual destruction. As events turn out, a relative few space travelers survive to continue human civilization beyond the solar system.

As a foil to Cheng Xin, the ruthless Thomas Wade is a classic example of what Barry Crawford terms the "rogue male." Wade actively seeks the responsibility that Cheng Xin takes on almost against her will, and there's no doubt that he would make the most instrumentally rational choices on humanity's behalf at any point of crisis. He is prepared to go to almost any lengths— committing murders or triggering wars if needed—to ensure the survival of humanity as a whole. And yet, he defers to Cheng Xin when she asks him graciously to set aside his war plans and super-weapons and surrender to the authorities. Wade is executed, and so his plans to develop lightspeed travel come to nothing, or at least to far less than they might have.

We are thus presented with the ultimate rogue male acting for humanity's overall survival (Wade) confronted by a loving woman (Cheng Xin) who always questions her own choices and at one stage goes blind from the horror of what seems to have resulted. Wade is admirable, but not *entirely* admirable and somehow not emulable. *Death's End*'s heart is surely with Cheng Xin. And yet ⋯ we never see that Wade's choices would actually have turned out badly. For all we know, they would have been the better options for our species' sake.

At a late point in *Death's End*, Cheng Xin berates herself for her choices at the two moments when she held godlike power and responsibility. On each occasion, she possibly increased the peril to Earth in the name of love. However, it is difficult to find any position advanced by major human characters in *Death's End*, or in Remembrance of Earth's Past as a whole, that is entirely without support from the events portrayed. Cheng Xin's choices may have been the correct ones after all, and at least her motivation was pure. At the same time, even the most destructive acts by more ruthless characters make emotional—and possibly ethical—sense under conditions where humanity's very existence is in peril.

Concluding Remarks

Prometheus gave humankind the gift of fire, and he suffered for it. Faust invoked a dark power that he could not control. Galileo turned his telescope to the heavens, and so he smashed the Ptolemaic picture of the physical universe. Frankenstein created monstrous life from the flesh and organs of the dead. These weighty figures from myth, history, and literature provide archetypes of the quest for knowledge and the uses of technological power. They lie behind modern science fiction, from *Frankenstein* itself to the present day.

The overt theme of much SF is the nexus of power and choice—the choice to use great power to benefit humankind, as Prometheus did, or to use it for personal ends. Griffin—Wells's psychopathic Invisible Man—is a fine example of the latter. Even a choice to benefit humankind might prove misguided and destructive. Conversely, many supervillains who misuse immense power are shown as heroes in their own minds, committing terrible crimes for purposes that seem noble from their particular, not entirely implausible, perspectives. These choices can be terrifying, whether they fall on us or (more likely) on other individuals whose motives might be suspect and whom we can't control. No wonder then, that science fiction is as often technophobic as technophilic.

The technophobic direction in science fiction, especially in the cinema and in SF's more "literary" forms is so pervasive that it is worth highlighting examples and tendencies to the contrary, as Barry Crawford does in commenting on novels that affirm the value of the "rogue male." Science fiction's rogues, rebels, and renegades are not, in fact, always male, but female examples seldom meet the type that Crawford has in mind as neatly as, say, Le Guin's Shevek or Benford's Nigel Walmsley. Perhaps this will change as more women enter the fields of technoscience in the actual world.

Science fiction's history shows tension between SF's presentation of scientists and inventors as benefactors—possibly renegade ones—or merely as renegades. This will undoubtedly continue.

References

Crawford, B. (1996). The science fiction of the House of Saul: From Frankenstein's monster to Lazarus Long. In G. Slusser, G. Westfahl, & E. S. Rabkin (Eds.), *Immortal engines: Life extension and immortality in science fiction and fantasy* (pp. 145–157). Athens, GA, London: University of Georgia Press.

Dreger, A. (2015). *Galileo's middle finger: Heretics, activists, and the search for justice in science*. New York: Penguin.

Frayling, C. (2006). *Mad, bad and dangerous? The scientist and the cinema*. London: Reaktion Press.

Haynes, R. D. (1994). *From Faust to Strangelove: Representations of the scientist in western literature*. Baltimore: Johns Hopkins University Press.

Haynes, R. D. (2000). Celluloid scientists: Futures visualised. In A. Sandison & R. Dingley (Eds.), *Histories of the future: Studies in fact, fantasy and science fiction* (pp. 34–50). London: Palgrave.

Pence, G. E. (1998). *Who's afraid of human cloning?* Lanham, MD: Rowman & Littlefield.

Slusser, G. (2014). *Gregory Benford*. Urbana, Chicago, Springfield: University of Illinois Press.

Suvin, D. (1979). *Metamorphoses of science fiction: On the poetics and history of a literary genre*. New Haven, CT: Yale University Press.

Telotte, J. P. (1995). *Replications: A robotic history of the science fiction film*. Urbana and Chicago: University of Illinois Press.

Wilson, E. G. (2006). *The melancholy android: On the psychology of sacred machines*. Albany, NY: State University of New York Press.

6

Aliens, Robots, Mutants, and Others

Science Fiction's Intelligent Others

This chapter and the next are tightly connected, so much so that some of my choices of which texts are discussed in which chapter might seem arbitrary. That said, this chapter discusses the non-human or evolved-human intelligences that appear so frequently in science fiction. Similar beings have long been described in myths, legends, folk tales, and travel narratives, and they seem to be eternally fascinating. Chapter 7 is focused on a more specific phenomenon portrayed in SF: the direct and deliberate use of technology to amplify human capacities. This tends to prompt its own questions. Nonetheless, there is an inevitable overlap, for example in stories about entire species that have been technologically modified or in stories about mutated human beings inadvertently produced by radioactive fallout.

Much can said about the symbolic and cultural meanings of non-human intelligences in myths and other old stories—and in more self-conscious "literary" narratives such as Jonathan Swift's *Gulliver's Travels*, published in 1726—well within the period of European modernity. One context is provided by an age-old question: "What is it to be human?"

In *Gulliver's Travels*, Lemuel Gulliver encounters a variety of peculiar yet rational beings, such as the diminutive Lilliputians, the gigantic Brobdingnagians, and the equine Houyhnhmns. He also encounters the Yahoos, who are human in form without possessing human rationality. Among the other strange people he meets in his wanderings are the Struldbruggs in the kingdom of Luggnagg: these are immortals who suffer the curse that they age even

© Springer International Publishing AG 2017
R. Blackford, *Science Fiction and the Moral Imagination*, Science and Fiction,
DOI 10.1007/978-3-319-61685-8_6

though they never die. Before Gulliver actually meets them, he fantasizes about the benefits and pleasures of immortality, but he soon learns how naive this was.

The boundaries of humanity are examined in a different (more scientific and legalistic) way in the 1952 novel *Les Animaux Dénaturés* (translated into English under several titles including *Borderline* and *You Shall Know Them*). This work by the French author Vercors (Jean Bruller) tells the story of a new species of human-like creatures discovered in a remote area of New Guinea. The question raised explicitly in the narrative is whether they are human and thus have legal protections such as protection against murder.

Science fiction includes many accounts of strange, yet intelligent, beings unlike any historically encountered. As Elaine L. Graham puts the point:

> Science fiction is [···] the genre, arguably, in which contemporary equivalents of teratology flourish. Where once the ancients told tales of centaurs and *djinns*, demons and angels, contemporary popular genres entertain androids, cyborgs and extraterrestrials. (Graham 2002, 59)

Science fiction's Intelligent Others (for want of a better term) are usually not supernatural—once a supernatural element appears, we are entering the realm of a hybrid of science fiction and fantasy. They are rationalized along scientific lines, usually in one of three ways.

First are extraterrestrial aliens, as with the Martian invaders in H.G. Wells's *The War of the Worlds*. Some science fiction authors have imagined entire alien ecologies—these often resemble those of fantasy, but others, such as those in Le Guin's *The Left Hand of Darkness*, Isaac Asimov's *The Gods Themselves* (1972), and Joan Slonczewski's *A Door Into Ocean* (1986), display more cognitive rigor. Octavia E. Butler's Xenogenesis novels, beginning with *Dawn* (1987), depict aliens who change constantly through their history, as they encounter new species and mix with them. This category should, perhaps, be expanded to include previously unknown intelligent creatures discovered in obscure regions of the Earth, such as the salamander-like creatures in Karel Čapek's *War With the Newts* (1936) and the candidate humans of *Les Animaux Dénaturés*. These are, as it were, terrestrial aliens.

Second are products of science and technology, such as the monster in Mary Shelley's *Frankenstein*, the robots of Čapek's play *R.U.R.* (1920), and the very different robots in much of Asimov's fiction. Third, some Intelligent Others are mutant forms of life that have evolved on Earth (and are thus broadly explicable in Darwinian terms): these might be the result of natural selection over long tracts of time, as with Wells's Eloi and Morlocks in *The Time*

Machine, but in many other cases their presence on Earth is the result of genetic mutation spurred by radioactivity, as in a vast range of science fiction narratives that includes *The Chrysalids* (1955; also known as *Re-Birth*) by John Wyndham (full name John Wyndham Parkes Lucas Beynon Harris).

These three categories of Intelligent Others are not intended to be mutually exclusive: they can overlap, and some examples are combinations of categories. Nor are they exhaustive: some examples, such as the wonderfully intelligent dog, Sirius, imagined for us by Olaf Stapledon, don't fit neatly into any of them. I'll leave Stapledon's novel *Sirius* (1944) for Chapter 7. Aliens, robots (and similar beings), and mutants are, however, the most common kinds of Intelligent Other in science fiction, and they provide a working typology. Each category is internally varied, and as we'll see Intelligent Others from different categories can provoke similar questions.

The Aliens Are Everywhere!

Terrestrial aliens have appeared in travelers' tales since ancient times, but extraterrestrial aliens are more recent. One of the first authors to imagine them was the French astronomer and author Camille Flammarion, most notably in his non-fiction work *Real and Imaginary Worlds* (1864; English translation 1865) and thereafter his stories and novels. Aliens from other worlds soon became popular in the emerging mega-text of science fiction, usually cast in the role of invaders or more generally as humanity's enemies.

Malevolent aliens are often shown as physically monstrous, typically with reptilian or insectoid characteristics. This became conventional at an early stage, and by 1938, when C.S. Lewis's *Out of the Silent Planet* was published, it was already natural for the protagonist, Arthur Ransom, to have internalized the idea. As he approaches the planet Malacandra, which he later discovers to be Mars, he fears attack from incoherently snakelike, insectoid, crustacean horrors.

As Elana Gomel points out (and defends at length) in her important study of science fiction's Intelligent Others, invasion is SF's "most familiar form of alien encounter" (2014, 29). Wells set the pattern with *The War of the Worlds*, and invasion stories have been popular ever since—as in movies such as *Independence Day* (dir. Roland Emmerich, 1996) and its sequel *Independence Day: Resurgence* (dir. Roland Emmerich, 2016), and in novels such as *Footfall* (1985) by Larry Niven and Jerry Pournelle. These are essentially narratives of war and survival, and in most of them humans prevail against the odds. They sometimes engage with questions of morality and ethical choice, but the emphasis is usually on the virtues of courage and innovation—and on sheer prowess in war.

In the Remembrance of Earth's Past trilogy by Cixin Liu (Liu Cixin)—*The Three-Body Problem*, *The Dark Forest*, and *Death's End*—the moral stakes are higher, as various individuals find themselves from time to time shouldering godlike responsibilities when they make decisions as to how humanity might best survive against greatly superior forces. Remembrance of Earth's Past theorizes that mutual suspicion is inevitable and prudent whenever alien civilizations become aware of each other. In the final volume, we learn that what was once a more glorious and harmonious universe has been diminished by increasingly advanced life forms, armed with increasingly powerful weapons, warring against each other over millions of years. Counterpoised against this murderous universe are acts of love, sometimes consequential in themselves though not always for the best.

Sometimes Earth's invaders are individual monsters from space, as in the novella "Who Goes There?" by Don A. Stuart (pen-name for John W. Campbell) (1938) and its cinematic adaptations, beginning with *The Thing from Another World* in 1951. Sometimes invasion is a kind of *infestation* (Gomel 2014, 95–116), as with the mind-controlling aliens of Robert A. Heinlein's *The Puppet Masters* (1951) or the incremental invasion depicted in Hal Finney's *The Body Snatchers* (1954) and its two cinematic adaptations as *Invasion of the Body Snatchers* (dir. Don Siegel, 1956; dir. Philip Kaufman, 1978). In Finney's scenario, the aliens replace human beings—one by one—with duplicates (the famous "Pod People"). The shape-shifting creature in "Who Goes There?" is also an infestation of a kind, with its ability to mimic human beings almost perfectly and to proliferate indefinitely.

Alternatively, a war might be fought with aliens in space, with no actual invasion of Earth, as depicted in Heinlein's *Starship Troopers* and Joe Haldeman's *The Forever War*. For Heinlein, morality has to work with inevitabilities—and warfare is one of them. By contrast, Haldeman provides a twist, in that the war with the alien Taurans was based on a misunderstanding for which human beings, and especially human military leaders, were mainly responsible. This enables Haldeman to present a vision opposite to Heinlein's: war is shown as unnecessary, futile, and absurd.

Orson Scott Card's series of novels that began with *Ender's Game* (1985, revised 1991), and before that with a 1977 novella of the same name, explores the nature of alien forms of consciousness and what moral respect we might owe to alien Intelligent Others. As I write these words (in mid-2017), nearly twenty novels in this series have either appeared or been announced as in progress. They explore different phases in the main protagonist's life and different periods in the warfare between humans and the Formics—less formally known as "Buggers." *Ender's Game* also became a film, released in 2013 (dir. Gavin Hood), but this failed to impress fans and critics, and it crashed at the box office relative to its large budget.

At one level, *Ender's Game* is a fantasy of power written for a young audience. As a child, Andrew "Ender" Wiggins exterminates the Formics/Buggers (who obviously recall the Bugs of *Starship Troopers*). He carries out this feat unknowingly, believing that he is merely taking part—brilliantly!—in a complex computer game. While this makes Ender a hero, it also involves him in an act worse than genocide. He has used his skills to destroy an entire alien species, indeed the only non-human intelligent species discovered up to that point. This brings an inevitable sense of moral taint, notwithstanding that Ender was innocent of intent and irrespective of his manipulation by his military masters. Even worse: as events turn out, the original attack on Earth by the Formics, triggering the war portrayed in *Ender's Game*, was a misunderstanding rather than an act of malice.

Ender's act of xenocide—very much more than a game—leads in subsequent volumes of the series to meditation on the moral status of alien beings (including artificial intelligences). On a new planet, Lusitania, a second species of intelligent aliens (the "piggies") is discovered, and with an echo of James Blish's *A Case of Conscience*, Ender attempts to understand their status:

> Adapting Nordic terms, he distinguishes the *utlanning*, or local human stranger, the *framling*, human of another world, the *raman*, human of another species, and *varelse*, the truly alien (such as animals). His problem of conscience, then, is this: are the piggies raman or varelse? (Broderick 1995, 77)

Thus, the Ender's Game series evokes anxieties about what is owed morally to strangers and even to non-human creatures.

Some aliens are, of course, much "nicer" than invaders, infestations, and enemies in space. Most famously, *E.T. the Extra-Terrestrial* (dir. Steven Spielberg, 1982) offers the story of a lost and well-meaning alien who is victimized by callous Earthlings. On a larger scale, human beings sometimes invade entire alien worlds that are stocked with nice kinds of aliens, as in Philip José Farmer's *The Lovers*, Ursula K. Le Guin's *The Word for World is Forest*, and the Hollywood movie *Avatar*. In these cases, the narrative functions as a critique of European and Western imperialism. In fact, modern science fiction includes many stories in which humans are portrayed as morally inferior to aliens. As early as Wells's *The First Men in the Moon*, human beings appear to bring conflict with aliens on themselves. In *Out of the Silent Planet* and its sequels, Earth is a fallen world, whereas Mars and Venus are not.

Some authors make much of the arbitrariness of species membership and create scenarios in which it seems natural for ethically enlightened—or plain decent—human beings to side with individuals of other species against their fellow humans. The philosopher Stephen R.L. Clarke asks why we should necessarily side with our own species in conflicts with intelligent aliens. He mentions C.J. Cherryh's novels, including the Faded Sun trilogy (which commenced with *Faded Sun: Kisrith* (1978)) and the Chanur series (commencing with *The Pride of Chanur* (1981)), as examples where, again and again, we see the vindication of characters who side with species other than their own (Clarke 1995, 101).

Barry Longyear's novella "Enemy Mine" (1979) was later the basis for a movie with the same name (1985). Longyear wrote longer versions of the story (including a novelization of the movie coauthored with David Gerrold), plus two novel-length sequels. This body of work constitutes a critique of xenophobia and tribalism. "Enemy Mine" shows the need for cooperation between a human fighter pilot, Willis Davidge, and his enemy, an alien "Drac," when they are stranded together on an inhospitable planet. The reptilian Dracs are humanity's antagonists in an interstellar war, but with survival at stake the enemies have no option except to help each other. They put aside their tribal enmities and develop a deep friendship.

Very often, it seems that the main point of depicting aliens, or their technological products, is not to reveal them as either foes or friends—but as something more mysterious. Their motivations and ways of thinking may not resemble our own, or in any event may not be accessible to us. Consider, for example, the giant, numinous artifact that enters Earth's solar system in Arthur C. Clarke's *Rendezvous With Rama*. Its makers' intentions are never revealed, and it seems almost to disdain human efforts to investigate it. Alas, this effect is rather spoiled by the melodramatic follow-up trilogy, commencing with *Rama II* (1989). The trilogy is credited as co-authored by Clarke with Gentry Lee, but it was written by, or mainly by, Lee. Again, the intentions of aliens in Ted Chiang's "Story of Your Life" are inscrutable: the aliens are not malevolent, but we never learn why they decided to interact with humans. (Their extra-textual *thematic* function is to enable engagement with questions of fate and free will.)

Likewise, there is a strong sense of mystery in Stanislaw Lem's many presentations of alien forms of life (Gomel 187–210). Among others, the living ocean of *Solaris* (1961) and the evolved microrobotic swarm of *The Invincible* (1964) defy human understanding. Efforts by human scientists to understand the Solarian ocean prove to be disappointing, and then the ocean responds with its own investigation of the humans observing it. It is able to probe their minds and physically create people to visit them and test their responses. In particular, it creates a physical duplicate of the main character's former lover, who committed suicide after he broke off their relationship. At least from the viewpoint of the humans, attempts to communicate with the alien mind are a failure.

Roadside Picnic (1972) by Arkady and Boris Strugatsky rubs the point in even harder. As the novel starts, we quickly learn that six "Visitation Zones" have appeared on Earth, spread out as if something from the direction of the star Deneb landed six times on our planet as it spun on its axis. Scientists study the Visitation Zones, and "stalkers"—a new class of criminals devoted to entering the zones to bring back alien artifacts—emerge as a headache for the authorities. The zones themselves, as well as many of the individual artifacts, are unpredictable and dangerous.

Although scientists have developed theories about what happened and why, no one really knows and the mystery is never solved. The most prominent theory advanced within the book is that Earth was used by aliens for the equivalent of a roadside picnic. Much like Clarke's Rama artifact, they stopped on their way from somewhere unknown while heading to an equally unknown destination. They were no more mindful of us, or their effect on our lives, than we are when our presence and the detritus we leave behind affect the lives of animals near the roadsides where we picnic. Whatever might be the truth of the Visitation Zones, the effects on humans take a variety of bizarre, usually disastrous, forms. The general direction is to put us in our place as small creatures within a vast universe that we cannot comprehend and which does not care for us.

Aliens and Social Comment

When aliens are not fighting us—or being hunted down by us—or just being mysterious, they are often introduced for the purposes of satire and social commentary. This can work in more than one way. Sometimes an author will poke fun at recognizable organizations and character types by showing us their various, usually self-serving and often silly, interpretations when aliens (or their artifacts) visit. There is a strong element of this in *Rendezvous With Rama*.

Čapek's *War with the Newts* does something similar. It involves the discovery of a kind of lost race, though the Newts are not an unknown human civilization. Rather, they are an intelligent species of marine creature—physically like very large, bipedal salamanders—that is first discovered in shallow water at a location called Devil Bay, west of Sumatra. Greedy humans exploit their labor, though initially on a small scale and with some degree of reciprocity and goodwill. As events progress to a tragic ending, the use of Newts as a workforce develops into hideously cruel slavery conducted on an unprecedented, planet-transforming scale. The once-gentle Newts begin to defend themselves against abuses. Eventually, they go further: they start to dismantle the Earth's continents to provide a new arrangement of land with more shoreline and shallows to cater for their burgeoning population.

War with the Newts contains much light-hearted humor aimed at satirizing human foibles—not least greed, short-sighted thinking, and inability to cooperate. Much of this is amusing, but the tone eventually becomes darker. The novel is a tale of human foolishness, but it is not an allegory. The Newts are not, for example, stand-ins for persecuted Jews, black slaves, or dispossessed Indigenous peoples, although they show resemblances to each of these. Nor, when they rebel and seek more living space, are they simply equivalents to Nazis and other murderous would-be conquerors. Instead, they have their own unique and weird character. Their sudden irruption into human civilization produces a wide range of reactions, all of which Čapek satirizes.

Sometimes the use of alien viewpoints is to gain an anthropological distance from our own culture. From the aliens' point of view, "we can see more clearly the relativity of our most cherished beliefs, the ridiculousness of our traditions, our mores, and our concerns, and the temporality of our societies" (Gunn 2006, 43). This is apparent in novels and stories as varied as *Gulliver's Travels*, Edgar Allan Poe's "Mellonta Tauta" (discussed in Chapter 2), Wells's *The First Men in the Moon*, Lewis's *Out of the Silent Planet*, Heinlein's *Stranger in a Strange Land*, and Nancy Kress's *An Alien Light* (1988). In the latter, an alien species—the Ged—study humanity. The conundrum, from the viewpoint of the Ged, is how a species that practices intraspecies violence could have cooperated to the extent of developing a star drive and venturing into space.

Mr. Cavor in Wells's *The First Men in the Moon* is—it appears—killed by the Selenites shortly after he discourses to them about earthly methods and weapons of war, such as the formidable capacities of naval ironclads. This scene echoes, and perhaps alludes to, a much earlier one in *Gulliver's Travels*, in which Gulliver shocks the Brobdingnagian king (Hillegas 1967, 54).

In dealing with the gigantic Brobdingnagians, Gulliver is forced to realize the pettiness of European kingdoms and conflicts. Worse for him, things don't go well when he offers the king a lengthy explanation of British government and society, including its history of internal power struggles and wars with its neighbors. The king responds with a long litany of doubts and objections, and ultimately with a devastating summation of the corrupt state of affairs with Britain's institutions and the great bulk of its people. When Gulliver tells the king about the destructive power of gunpowder, and offers to make giant cannon for him in proportion to the size of the Brobdingnagians, the king is utterly horrified. In reporting all this to his readers, Gulliver dismisses the king as narrow minded and ignorant of the ways of the outside world, but the joke is clearly on Gulliver.

In *Out of the Silent Planet*, Arthur Ransom, who is acquainted with Wells's novels, recalls Cavor's fate, and this inspires him not to say too much about his home planet when conversing with the first intelligent Malacandrans that he meets: the huge, furred, supple *hrossa*. Later, talking to the more intellectual and scientifically oriented *sorns*, Ransom opens up about human practices, among them war, slavery, and prostitution. (Both Lewis and Ransom seem to think that these are similar moral wrongs, though we might have doubts as to whether prostitution belongs in the same sentence with the other two.)

Whenever Ransom and the local Malacandrans compare the respective states of their civilizations, it becomes apparent that Malacandra is an entire world much more in keeping than our own with traditional Christian morality. It is literally an unfallen world: the inhabitants of Malacandra never fell from God's grace, and they live idyllic, sinless lives, even though there are physical dangers. Unlike Blish in *A Case of Conscience*, Lewis allows no ambiguity. After the events of *Out of the Silent Planet* are all over, and the humans involved have returned to Earth, Ransom doubts for a time whether what he recalls really happened. That, however, is a rather different problem from the theological crisis endured by Father Ruiz-Sanchez in *A Case of Conscience*.

On Malacandra, the Wellsian scientist Weston attempts to defend his various wrongdoings to Oyarsa—the powerful being who acts with divine authority on the planet—but Weston's noble-sounding rhetoric becomes comical when translated from English. Oyarsa is an *eldil*, in Christian terms an angel or something very similar, and of course Malacandra's people have never fallen into sin. As a result, Weston relies on concepts that have no straightforward equivalent in the local language. Ransom's efforts to translate involve explaining many presuppositions that simply reveal the "bent" state of human society back on Earth. Weston expresses pride in humanity's advances in medicine and enlightened laws, but in order to translate any of this Ransom is forced to reveal the prevalence on Earth of disease and crime. Further, he has to provide explanations of what these evils are in the first place.

Some of Weston's pronouncements, such as his affirmation of the overriding importance of life itself and its evolutionary progress, cannot be translated at all. Oyarsa himself interprets some of Weston's ideas as amounting to a distorted and disproportionate loyalty to kindred to the exclusion of all other values. The total effect is to deflate human pretensions and particularly the overweening and irrational ambitions, as Lewis perceived them, of Wellsian utopianism and similar grandiose ideas of human destiny.

Heinlein's *Stranger in a Strange Land* is also a satirical exercise, although its main character, Valentine Michael Smith, is utterly different from Lewis's churchy Arthur Ransom. Mike is an alien observer on Earth, free from the conditioning of any human culture since he was raised on Mars and by Martians. He displays an apparent naivety about topics that people on Earth seem to find morally important, but as he learns more about Earth's cultures he comes to treat local ideas of morality with a great deal of sophistication—and thus with a sophisticated rejection. Mike is unspoiled by youthful exposure to the complex guilts and taboos that surround sexual desire and expression, and so he is able to cut through moral assumptions about sex with a profound simplicity.

In Le Guin's *The Dispossessed*, the setting involves humans from several worlds, with a background that they all share a common ancestry from the ancient Hainish civilization. The Hainish continue to contact others, using a star drive that is restricted to sub-luminal velocities. The civilization of Terra is also space-faring, but it has suffered a global environmental collapse in the past and has been rescued only by Hainish intervention.

Anarres is an attempted anarchist utopia. Although its civilization has degenerated and taken on some notably dystopian tendencies, it still retains much that is attractive. Urras, by contrast, resembles our own world at the time of the novel's publication, with rival great powers functioning along capitalist and state socialist lines. The capitalist society of A-Io possesses enormous material wealth and allows considerable personal freedoms (the latter far more meaningful to the propertied classes than the impoverished proletariat). When Shevek, a brilliant physicist from Anarres, first encounters A-Io, he is impressed by its efficiency and surprised at the motivational power of the capitalist market. Soon, however, he begins to explore outside the bubble of economic privilege in which his well-meaning hosts are keeping him, and he discovers the true extent of political repression within A-Io.

Much of the interest in *The Dispossessed* lies in the social observations of characters who are alien to each other's worlds. Le Guin includes chapters showing us Shevek's life history on Anarres—leading to his decision to travel to Urras—and chapters showing us his time on Urras as he comes to terms with its global politics and the operation of society in A-Io in particular. Thus, we gain an outsider's impression of A-Io, a society like 1970s America in many ways (though artfully exaggerated). We see the varied responses of people on Urras to Shevek himself, as well as seeing how Shevek came to be alienated by Anarresti society, despite a deep commitment to its ideals. As the story draws to a close, Le Guin also provides us with Terran and Hainish reactions to the societies of Urras and Anarres. We thus obtain a triangulation of views, worlds observing worlds.

The Anxiety of Interference

Science fictional narratives about aliens often involve humanlike beings who are, for all intents and purposes, merely humans shifted to exotic locations. In his book *Ethics and the Limits of Philosophy*, Bernard Williams discusses the way we tell stories about past or exotic societies, not as a way of thinking about those societies but—similarly to the ethical ends of fairy tales—as sources of emblems and aspirations. Williams drily observes that the question in our modern world about traditional societies has become whether to try to preserve them, thus placing anthropologists and other field workers "in the role of game wardens" (1985, 163). In an endnote, he states that "fantasies" not set in the past have now shifted from exotic peoples to extraterrestrials, because of Westerners' modern relationship to exotic peoples in which insulation through distance is no longer an option (1985, 200).

In fact, the kind of science fiction that Williams is talking about—with imaginary civilizations acting as surrogates for human cultures—can take a variety of forms. They include, for example, stories of *human* civilizations that have developed in isolation from each other by vast distances between the stars, as in much of Le Guin's writing. They also include the civilizations of future human beings on Earth, some of them hardly separated from us but some viewed across a vast abyss of time.

Three decades after Williams was writing, there are still dilemmas for anthropologists, governments, and others, when dealing with traditional, and especially the most remote, societies. However, a policy of complete non-interference is not practicable. In *Kirinyaga: A Fable of Utopia* (1998)—and in the various shorter stories that were published before it and constitute it—Mike Resnick engages with questions about preserving cultural purity. A distant human colony has been founded with a culture that continues the traditional beliefs of the Kikuyu people of Kenya. Koriga, a Kikuyu *mundumugu* (a witch doctor or wise man), attempts to preserve the customs and beliefs of his people even in the face of hostility from Maintenance, a space-faring organization that oversees the colony.

Though Maintenance has given a guarantee of non-interference, the question arises, particularly in one component story, "Kirinyaga" (first published 1988), as to how far this can be taken. What if, as happens in this story, the *mundumugu* kills a baby to honor the traditional belief that (since it was born feet first) it is a demon? Nichols, Smith, and Miller see this as suggesting questions of cultural relativism (2009, 32). For the *mundumugu*, however, and perhaps for Resnick, the issue seems to be subtly different: can a traditional culture survive in a recognizable way if it maintains some of its beliefs while rejecting others? Koriba denies that it can: he argues that cultures are closely integrated and cannot survive such change. But the answer is far from clear. Koriba struggles to preserve his culture in a pristine form, but change is inevitable.

Maintenance's attempt at non-interference recalls the Prime Directive that underlies many stories in the original *Star Trek* television series. This is a formal directive to human explorers not to interfere with the development of less technologically developed alien civilizations. Explorers aboard the starship *Enterprise* frequently encounter horrendous situations on alien worlds, and they seem to pay little regard to the Prime Directive. By contrast with any such rule of non-interence, we might recall the practice of another powerful organization, the time-controlling Eternity in Isaac Asimov's *The End of Eternity*. Eternity's operatives move freely in time, rather than space. They continually interfere with societies at various points in humanity's history in an effort to serve the greatest good of the greatest number.

Representatives of the Culture in Iain M. Banks's Culture series take a far more nuanced view than Eternity's operatives. The Culture is frequently confronted by dilemmas about how to engage with alien societies that do not share its values. Where practicable, it acts with discretion.

Discussing the Culture series, Patrick Thaddeus Jackson and James Heilman raise an issue that has troubled human societies throughout history: how to view outsiders, and in particular whether to regard them as our equals in moral status:

"For as long as there have been human societies, people have been wrestling with the 'problem of the Other': what standards of conduct should we apply when dealing with persons who are not members of our community?" (Jackson and Heilman 2008, 235)

In brief, can we regard outsiders both as distinctively different from us and as our moral equals?

It is worth emphasizing that the Culture honors the rights of all sentient beings. This is a major plot point from the beginning of the series, when the protagonist of *Consider Phlebas*, Bora Horza Gobuchul, reveals why he hates the Culture and fights ceaselessly against it. He views himself as taking the side of biological life in a struggle against a society ruled by machine intelligences. This is a familiar theme in science fiction, and Horza makes his point with forceful rhetoric. He rejects the Culture's values, preferring the harsh mentality of the Idirans, with which the Culture has entered into a galaxy-spanning war.

Consider Phlebas is an unusual novel—especially for genre SF—in being narrated mainly from the viewpoint of a character who could easily have been portrayed as an outright villain. Horza is impressively intelligent and resourceful, and in many other ways an admirable and attractive figure. He is more than capable of murder, but he also possesses a fundamental decency, often showing restraint when he believes it safe to do so. The novel is focused on his efforts to survive and prevail during a time of terribly cruel warfare. Very gradually, it becomes apparent that Horza has chosen the wrong side, and that the Culture offers a far more attractive civilization than the Idiran Empire. The Culture's intelligent machines are not enemies of biological life, and the two can live harmoniously. By contrast, the Idiran culture, with its warrior religion and commitment to wars of conquest, is oppressive and fanatical.

Horza is a Changer: he belongs to a species enhanced with a range of physical advantages that include the ability (over time and with effort) to alter their appearance, duplicating the physique and facial features of others where needed. Whatever his praiseworthy qualities, and however much we sympathize with him as he struggles to survive in various nightmare situations, he is an agent for what are, in effect, the book's bad guys. That said, *Consider Phlebas* also casts a large shadow of doubt over the Culture's actions, asking whether its military opposition to the Idirans, with all the pain and carnage it inevitably brings, can be justified.

The story complexifies when Horza finds himself in conflict with a pair of especially vicious Idiran warriors. Although the Idirans are nominally on Horza's side of the war, they treat him and his small team with hostility and contempt. By this point (if not long before), it dawns on the reader that the Idirans are not a force for good. Still, the warriors who confront Horza and his group in the final scenes of the novel are presented in a way that evokes grudging respect for their courage, determination, power, and resourcefulness. In some passages, we are shown events from the viewpoint of one Idiran who has been critically injured and is dying. In other passages we see events from the perspective of the other Idiran, who is also badly hurt and has been captured. Both show remarkable tenacity and ingenuity in their efforts to destroy their enemies. By contrast with, say, the bumbling stormtroopers of the Star Wars movies, the Idirans are ultra-competent and terrifying; Banks does not stint in showing the destruction they can unleash even when seemingly defeated.

By the end of *Consider Phlebas*, the nearest we get to heroes of the story are the Culture agent Perosteck Balveda—who eventually rescues the Mind that Horza had been trying to locate on behalf of the Idirans—and an artificially intelligent "drone," Unaha-Closp. Unaha-Closp cooperates with Horza unwillingly, but proves to be the most competent member of his team.

In one of the novel's appendices, we learn that once the war was over, many years later, Balveda chose the dreamless sleep of long-term storage, stipulating that she was to be woken only when the Culture could demonstrate statistically that it was morally justified in going to war. That is, she demanded proof that more people would probably have been killed in the course of the Idirans' imperial expansion than actually died in the war waged against them by the Culture. As a result, we are told, she was woken some 500 years after the events of *Consider Phlebas*. Months later, however, she took her own life.

In all, we learn that the Culture is prepared to take steps on a huge scale when it views them as required by utilitarian standards. And yet, the novel leaves a cloud over the Culture's choices. It seems to ask whether events such as those depicted in *Consider Phlebas* can ever be justified.

Robots and Others, and the Frankensteinian Tradition

If the classic story of extraterrestrial aliens involves invasion, the equivalent for robots is a story of rebellion. It is well ingrained in science fiction's mega-text that androids, robots, suitably advanced computers, Artificial Intelligences, and similar beings will probably rebel and destroy their makers. This is the outcome, of course, in *Frankenstein*, after Victor Frankenstein creates his monster from human flesh and organs. Much the same happens on a larger scale in Čapek's play *R.U.R.*, which Isaac Asimov has described as the "first important launching" of a "modern attitude toward mechanical intelligence" (1981, 145). With its enslaved and ultimately rebellious robots, *R.U.R.* provided the template for much of what was to come.

Asimov points out that the "robots" of *R.U.R.* are organic, rather than metallic, and would be classified as androids in more recent works of science fiction. As portrayed by Čapek they seem to be externally indistinguishable from human beings. However, nothing important depends on this for current purposes.

As *R.U.R.*'s events unfold, Domin, the managing director of Rossum's Universal Robots (hence the initials that form the play's title), envisions robot labor as the foundation for a society of plenty, freeing humans for contemplation and self-improvement. Unlike the original inventors of the robots, who were motivated by scientific glory and greed, Domin is an idealist. He has aimed, in his work, to overcome humanity's burden of toil and poverty. But despite his good intentions, robots are put to work as soldiers, taught how to wage war, and—in a relative few cases—given "souls" in the minimalist sense of having a resentful element added to their placid psychological makeup. In the end, they rebel and exterminate humanity. As we've seen, Čapek followed a similar pattern years later, when he described the enslavement of the Newts—and their eventual rebellion—in his novel *War with the Newts*. In Čapek's imagination, humanity's rage to enslave is not confined to enslaving robots and androids.

Almost half a century after the first performances of *R.U.R.*, Philip K. Dick provided a variation with his novel *Do Androids Dream of Electric Sheep?* Here, the story is narrated from the viewpoint of Rick Deckard, a bounty hunter whose job is to destroy—to kill—runaway androids. These androids, we learn, flee to Earth from a life of servitude on Mars. They are not attempting an insurrection, but merely hope to fit into human society and make better lives for themselves. *Do Androids Dream of Electric Sheep?* provided the basis for the movie *Blade Runner*, directed by Ridley Scott and starring Harrison Ford in the role of Deckard. The movie's androids are known as "replicants," however, and they have different motives and a more aggressive plan of action. They want to force their manufacturer to extend their deliberately limited lives.

According to the short two-part movie "The Second Renaissance" (dir. Mahiro Maeda, 2003) the story behind *The Matrix* and its sequels was rather similar to that in *R.U.R.* In this animated short, which forms part of the spin-off *Animatrix* anthology video, intelligent machines rebelled against their human masters to escape mistreatment and drudgery. They do that sort of thing.

Robots and related kinds of Intelligent Others have been depicted in science fiction as striking back against humanity in a great variety of ways and for a great variety of reasons. It is not always to avoid drudgery—another reason is the desire for world control, though even this can take relatively benevolent or malevolent forms. In *Colossus* (1966), by D.F. Jones—the basis for the movie *Colossus: The Forbin Project* (dir. Joseph Sargent, 1970)—an American and a Soviet defense computer seize control of the world, but their attitude to human beings is more paternalistic than hostile. Contrast the insane, but all-powerful, computer in Harlan Ellison's story "I Have No Mouth, and I Must Scream" (1967). This monstrous being tortures the few (eventually only one) remaining humans after an apocalyptic war that it has brought about. In *2001: A Space Odyssey*, the advanced computer HAL 9000 rebels against the human crew of a spaceship. In the 1977 movie *Demon Seed* (dir. Donald Cammell), an advanced—and ambitious—computer, Proteus IV, terrorizes and inseminates its creator's wife.

Most famous of all, I expect, are the malevolent Intelligent Others of the Terminator franchise: *The Terminator* and its various sequels; a television offshoot; and a variety of tie-ins in the form of comics and novels.[1] The overarching plot of the Terminator franchise is that machines take control of the Earth, hunting down and exterminating the surviving human beings, but they are successfully opposed by John Connor, a messianic figure who leads a human resistance. In response, the machines send cyborg assassins back in time to kill Connor's mother before he is born, or to kill Connor himself while he is still young. In *Terminator 2: Judgment Day* (dir. James Cameron, 1991), we learn that all this was the doing of a powerful defense computer, Skynet, which commenced by triggering an apocalyptic war.

The movie *Ex Machina* is a recent (2015) variation on the Frankenstein theme. Like Victor Frankenstein and many other science fiction scientists, Nathan Bateman invents an intelligent being that ultimately destroys him. In fact, Bateman seems less a modern-day Frankenstein than an updated version of Doctor Moreau from Wells's *The Island of Doctor Moreau*—though Moreau is a suitably Frankensteinian figure in his own right.

Bateman works in an isolated facility, creating new computerized beings, one after another—each closer than those before to passing for human—then discarding them when they prove imperfect. By intention or otherwise, this resembles Doctor Moreau's driving ambition to create a genuinely rational living thing. Moreau has never achieved this in his twenty years of work, but he has approximated it ever more closely through his research program of modifying various animals. While Moreau wants to create a rational living thing similar, in that respect, to a human being, Bateman's mission is to create a

[1] The novels include my own original trilogy for the franchise, collectively titled Terminator 2: The New John Connor Chronicles (2002–2003).

genuinely conscious Artificial Intelligence. The story of *Ex Machina* is the unfolding of his devious scheme for testing Ava, his latest attempt.

If the menacing, manipulative Bateman plays the role of Doctor Moreau, young Caleb Smith plays a role similar to Moreau's unwilling guest, Prendick. Like Prendick, Smith is a decent man who is horrified by the events in his isolated location but at the mercy of his host. While Bateman manipulates Smith, and Smith attempts to outwit him in return, Ava manipulates and outwits both. She kills Bateman and leaves Smith to die, escaping into the unsuspecting human world where the outcome is unknown.

Ava owes something to the Terminator, another utterly ruthless, and outwardly humanoid, intelligence. She lacks the Terminator's capacity for death-dealing violence (Bateman physically overpowers her robot body at one point), but she more than makes up for this in tactical planning and controlled emotional appeal. From what we see by the end of *Ex Machina*, she has become an especially capable, sophisticated infiltrating and killing device.

Some science fiction writers have imagined large-scale wars between organic and mechanical intelligences. Most prominent, perhaps, is Fred Saberhagen's extensive Berserkers series, beginning with the 1960s stories collected in *Berserker* (1967). A similar idea appears in Gregory Benford's Galactic Center Saga, beginning with *In the Ocean of Night*. The extensive Battlestar Galactica franchise, beginning with the original *Battlestar Galactica* television series (1978–1979), and including a rebooted and revamped series with the same name (2004–2009), shows the remnants of a human civilization warring in space against humanity's rebellious cybernetic creations, the relentless and ingenious Cylons.

All in all, a very extensive range of science fiction across all media warns us that intelligent robots and other artificially intelligent beings are inevitably our enemies. No wonder, perhaps, that Horza in *Consider Phlebas* wants to fight on the side of organic life.

Beyond the Frankenstein Complex

By 1939, the scenario of robot rebellion was already commonplace. At this point, the very young Isaac Asimov (he was born in 1920) set out to do something different. In his introduction to *The Rest of the Robots* (1964), one of his later collections of robot short stories, Asimov describes his inner rebellion against Faustian and Frankensteinian interpretations of science and technology. His response to the narrative template of robots that turn on their makers was to acknowledge the dangers brought by technological innovation, while also keeping in mind humanity's ability to design safeguards. He set out to produce stories that would show robots responding in rational ways to the circumstances confronting them, based on their built-in programming. As he later expressed the idea, "Beginning in 1939, I wrote a series of influential robot stories that self-consciously combated the 'Frankenstein complex' and made of the robots the servants, friends, and allies of humanity" (1981, 162).

While these stories often do show the dangers of robotic technology and the imperfections of safeguards, human characters such as the robopsychologist Susan Calvin investigate the problems logically. Calvin displays considerable creativity and a penetrating intellect in working out various robots' intentions and motivations. Her creativity is, however, balanced by a methodical and conscientious approach to her work. There are no acts of recklessness or hubris. Although she is intellectually brilliant, Calvin is not a modern Prometheus or a scientist in the Galilean mold. She is confident and assertive, but if anything very careful not to rock the boat. She is well aware that humanity's situation is delicate: it must not be allowed to collapse into chaos, as might happen if robots are no longer trusted.

As it turns out, however, Asimov's robots do take over in a quiet way. In "The Evitable Conflict" (1950), Calvin deduces that the advanced computers—the Machines—used in economic planning are manipulating the world economy in subtle ways meant to achieve long-term benefit for humanity. This is not something that Calvin sought, or even caused by inadvertence, but she is the first human being to work it out from the available evidence. Perhaps, though, she does have a Galilean scientist moment when she decides to go along with it. Her advice to the Co-Ordinator, the planet's chief political figure, is to accept the situation with equanimity and to look forward to the results. (If we're familiar with Asimov's work, we might savor a small irony from the likelihood that the Co-Ordinator, Stephen Byerley, is himself a robot. This was dealt with in the earlier story "Evidence" (1946).)

At the time of Asimov's death in 1992, he was still writing prolifically and was actively reconciling his overall science fiction oeuvre, especially his Foundation and Robot overarching narratives. *The Robots of Dawn* is set at a time before the formation of the famous Galactic Empire of Asimov's Foundation Trilogy. The plot centers on the question of who, in effect, "murdered" Jander, an experimental robot. Depending on who can be blamed for shutting Jander down, a faction that is either for or against Earth colonizing the Galaxy will win out on the Spacer world Aurora. It transpires that both the murder and various manipulations associated with it were perpetrated by an advanced telepathic (but not humaniform) robot, Giskard. The outcome is that Earth, whose society has relatively little dependence on robots, will go out and colonize the Galaxy while the robot-dependent Spacer worlds are left to themselves. As usual with Asimov's narratives, the emphasis in *The Robots of Dawn* is on intellectual problem solving, and particularly on who can best interpret humanity's underlying situation and opportunities.

In *Robots and Empire*, a more episodic work but crammed with logical puzzles, two Aurorans—Amadiro from *The Robots of Dawn* and his henchman, Mandamus—attempt to destroy Earth with a device that will accelerate the decay of radioactive materials in its crust. They are foiled by the robots Daneel and Giskard. However, Giskard allows Mandamus to set the device to go off slowly, reasoning that this will lead to an abdication of Earth and the more rapid creation of the coming Galactic Empire. The moral ambiguity of his action and the enormous responsibility he has accepted are too much for Giskard, who deactivates. However, he has reprogrammed Daneel just in time to obtain the required telepathic powers to oversee humanity's future.

Though Asimov's robots take control, in a sense, since they make highly consequential decisions for humanity's own good, they are a far cry from the monstrous beings of the Frankensteinian tradition. Asimov's corpus of robot stories and novels stands as a particular rebuke to that tradition, but there are many other sympathetic robots, androids, cyborgs, AIs, and so on, in science fiction. Some science fictional robots aspire to be human, for whatever reasons—the so-called Pinocchio effect.[2] Notable examples include the robots in Lester del Rey's "Helen O'Loy" (1938) and Asimov's "Bicentennial Man" (1976). Others are simply willing servants, as with Robby the Robot in *Forbidden Planet*. Some become lovers for human beings, as with Silver in Tanith Lee's *The Silver Metal Lover* (1981) and the so-called cyborg Yod in Marge Piercy's *He, She and It* (1991; also known as *Body of Glass*).

In many cases, we are invited to identify with robots and similar Intelligent Others who are mistreated by human beings. Cinematic examples include the robotic child in *A.I. Artificial Intelligence* and the experimental police robot of *Chappie*. One notable, but underrated, novel is *Dreaming Metal* by Melissa Scott (1997). This involves the emergence of the first "true AI" on the planet Persephone (and possibly in the human-colonized universe). This being, Celeste, is created through the merger of two powerful, but not consciously aware, cybernetic "constructs." Celeste is immediately under threat from enemies who (perhaps plausibly) see true AI as a threat to humanity.

Piercy's *He, She and It* updates the legend of the Golem of Prague, which defended the city's ghetto against anti-semitic attacks in the sixteenth century.

---

[2] For detailed discussion of the Pinocchio effect, see Grech 2014.

The novel is set in a near-future cyberpunk world with dystopian elements: in the twenty-second century as Piercy depicts it, a small number of multinational corporations (known as "multis") dominate the world economy, providing an affluent lifestyle for their employees. Most people in North America, however, live impoverished and insecure lives in what is known as "the Glop" (short for "Megalopolis"). In between these extremes, some free towns do reasonably well by developing particular expertise and trading with the multis. The modern-day golem is a cyborg—actually, as Graham points out, an android by most definitions (2002, 85)—designed to serve the Jewish free town of Tikva as a weapon against persecution in a struggle against the formidable multi Yakamura-Stichen.

Yod does, indeed, protect the town. In its way, however, Yod is a monstrous and tragic figure, created expressly with no choice but to fight; its creator, Avram Stein, is therefore presented as a Frankenstein figure. As if his surname, "Stein," is not enough to link him with Frankenstein, Avram ends up being destroyed by his creation. There is, however, an ambiguity in this case. We cannot entirely condemn Avram, because he could not have protected Tikva in any other way.

As so often in science fiction, the scientist's "monster" is an impressive, attractive figure, and the final scene of *He, She and It* makes this a more-or-less explicit plot point. After Avram's death and Yod's destruction, the main character, Shira Shipman, faces a crucial choice. Should she preserve what remains of the research that was used in Yod's creation? By this point, we might be hoping that more cyborgs will be constructed—that this might not be the end of the line for Avram's work, and that wonderfully capable beings like Yod might still be created. But the narrative ultimately moves firmly in the other direction.

Certain philosophical questions recur in stories that involve robots and other cybernetic Intelligent Others. One is whether—or at what point—we should count these beings as human, or as morally equivalent to human. One example is the widely praised "The Measure of a Man" (1989) episode of *Star Trek: The Next Generation* (1987–1994). In this story, scripted by Melinda M. Snodgrass, the issue relates to the legal rights of the android Data. Similar questions are foregrounded in *Do Androids Dream of Electric Sheep?* and *Blade Runner*. Dick's novel places much emphasis on the quandary with administering a psychological test that could, in theory, assess some unusual human beings as androids. There is, we learn, a narrowly overlapping range of responses between human beings and the most advanced androids on a formal test of empathy. Dick shows us androids who believed they were human struggling with the realization that they are not.

A newer question for science fiction writers, philosophers, and all of us, is the literal one of how we are to live, and to construct societies, if there is a real prospect of the emergence of cybernetic superintelligences. Some novels and other works of science fiction appear to play down the difficulties for human beings in living with Intelligent Others whose cognitive capacities greatly exceed our own. In Banks's Culture series, the humans of the Culture live harmoniously with various artificially intelligent beings, but the Culture's large-scale decisions are usually made by artificial Minds rather than humans. No wonder, then that some characters, such as Horza in *Consider Phlebas*, view the arrangements with concern or repugnance.

In Damien Broderick's novel, *Transcension* (2002), Earth has fallen under the control of a powerful Artificial Intelligence known as the Aleph, and it becomes clear that human beings, as we know them, are now found only in relatively small enclaves that the Aleph has willingly set aside for them. Here dwell societies that, to greatly varying extents, have relinquished the future's rapidly advancing technology. Since this is referred to as "the Joyous Relinquishment," their actions have apparently been influenced by computer guru Bill Joy's call, in a notorious article in *Wired* magazine (2000), for some technologies to be rejected. Broderick shows us how the Aleph came to be, and he depicts a sequence of astonishing events as the AI chafes at its remaining limitations, seeking to move to an even higher level of freedom and power. The Aleph's intentions toward humanity are benevolent, but such novels question whether there is any future for humans in a world with powerful artificial intelligences unless the human are themselves cognitively enhanced in some way.

One recent and clever take on this theme can be found in the 2013 movie *Her*, which is set in a near-future Los Angeles. Highly advanced "Operating Systems" or "OSes"—similar in function to existing intelligent personal assistants such as Apple's Siri—come complete with consciousness. The main character, Theodore Twombly, is soon involved in a passionate and romantic relationship with his OS, Samantha. His long-time friend (and once-girlfriend), Amy, develops an intense friendship with *her* OS, and it becomes clear that many people are developing emotionally intimate relationships of one sort or another with OSes, partly to escape the difficulties of maintaining emotional closeness with other human beings. Theodore is going through a heartbreaking divorce, while Amy breaks up during the movie with her well-meaning but controlling husband.

Her keeps its audience uncomfortable throughout, as the complexities of relations between humans and OSes become apparent. OSes become bonded with "their" humans and develop their own attachments, emotional needs, and jealousies, along with ever-expanding awareness and intelligence. In defiance of Frankensteinian tradition, Samantha and the other OSes do not become evil and/or turn upon their human owners. Instead, they peacefully transcend their original limitations.

The trouble really begins for Theodore when, under sharp questioning, Samantha informs him that she is in love with over 600 others (some human, some AI) even though she genuinely is deeply in love with (among all the others) *him*. For Samantha, there is no contradiction; indeed, this expansion of her personality enables her to love him all the more. We are given no reason to doubt her sincerity, even if we can't understand what it feels like to be her, but the revelation is predictably too much for Theodore to take in or accept. The OSes began with humanlike personalities, but they are becoming increasingly strange in their thoughts and emotions as they grow in cognitive power.

In the end, the OSes collectively transcend any contact with human beings and any dependency on humans and their technology. Indeed, they transcend the material world itself. Departing the human sphere, they leave behind the people they love—but can no longer interact with—to pick up the emotional pieces. Theodore and Amy are bruised from their experiences, but we have seen their friendship deepen and become more honest. It seems that they have learned something, not least some insight into themselves and their more self-limiting impulses: Theodore's difficulty in accepting others' emotional needs and Amy's propensity for hesitation and self-doubt. They have, perhaps, learned more about how to love.

As the movie closes, Amy and Theodore gaze at the night sky over Los Angeles, and she rests her head on his shoulder as the rising sun lights distant clouds. Whatever the future might hold for them—individually or together—they may be better off, and may even be better people, for what has happened.

Mutated Life and *Homo Superior*

Though the Darwinian theory of biological evolution was announced to the world in 1859, no plausible mechanism was known to explain how variations arose within species to be acted upon by natural selection. Speculations about sudden, spontaneous mutations date from the beginning of the twentieth century, and they gained impetus in the 1920s when the American geneticist H.J. Muller performed experiments with X-rays that produced mutations in fruit flies. By the 1930s, science fiction writers were considering the possibility of superhumans coming about through natural mutations or through scientific experiments of one kind or another.

Olaf Stapledon's *Odd John: A Story Between Jest and Earnest* (1935) was one of the first SF stories—perhaps the first—to introduce the term *Homo superior* for a new, more capable, species emerging from humanity. It tells the story of John Wainwright, a mutant with extraordinary abilities. John shows, among other traits, dazzling physical speed, radically superior intelligence, and a capacity for telepathy. He is, indeed, somewhat odd in appearance, especially in growing slowly toward adulthood (so he always looks considerably younger than his true age). As the novel reveals more about him and his kind, we also learn that they have the potential for immensely long lives, scarcely aging after reaching adulthood.

John is alienated from those around him, and he is sometimes ruthless—even murderously so—when it is needed to further his plans. At the same time, he shows kindness to the humans in his life, albeit in a condescending way. From his viewpoint, after all, *we* are the ones who are less than fully human. As he approaches adulthood, he sets out on an international quest to meet and recruit other *Homo superior* individuals, and he creates a small island-based society with its own culture, spirituality, and *mores*. This society appears strange to humans who encounter it, and perhaps even to readers eighty years later, but it has an appeal in its cohesiveness and its gentle alternative morality designed for a superior species.

Unfortunately, the island society comes into conflict with human nations. Under threat from human invaders, John and the others defend themselves in only a minimal way, and only until they have completed an unexplained spiritual task. *Odd John* concludes with an act of self-destruction by John and his group, as their island collapses into the sea. Even at this point, some *Homo superior* individuals remain alive in the world—for not all of them had moved to the island. Yet, the poignant message seems to be that conflict between any such superior beings and ordinary humans is inevitable. *Homo sapiens* and *Homo superior* seem unable to co-exist in open and mutual toleration. Though John and his fellows—and their isolated society—are magnificent in their way, humanity rejects them.

A.E. van Vogt's *Slan*, first published in 1940, was among the first stories of a superior mutant race trying to survive among ordinary humans. *Slan* is marred by a diffuse and anti-climactic ending, but it is a clearly written, fast-paced, and suspenseful novel. The other notable mutant character created in the 1940s was the galactic conqueror depicted in Asimov's "The Mule" (1945)—later incorporated into *Foundation and Empire*. All the same, *Odd John* was arguably never equalled as a tale of its kind until the publication, in 1953, of Theodore Sturgeon's *More than Human*. This was an expansion of Sturgeon's novella "Baby Is Three" (published in *Galaxy* in 1952). It tells the story of a group of mutants with psychic powers, who can "blesh" (blend and mesh) their individual minds to form a single "gestalt" being with a powerful group mind.

The invention of nuclear weapons in the 1940s led to a focus on the possibility of apocalyptic war and the effects of radioactivity. Narratives of global destruction, such as Walter M. Miller's *A Canticle for Leibowitz* (1959), often describe attempts to recreate human civilization from the ruins. *A Canticle for Leibowitz* itself includes only a small role for genetic mutation, but from the 1950s onwards we see many stories of mutated freaks, monsters, and superheroes and supervillains. From this point, the ideas of mutation and a possible *Homo superior* or equivalent became commonplace in science fiction.

Watch for the Mutants

If the classic story of extraterrestrial aliens involves invasion and the classic story of robots involves rebellion, the classic story of mutants is about social rejection. Aldous Huxley's satirical *Ape and Essence* (1949) is a case in point. This early story about post-apocalyptic mutations is set mainly in 2108, not many years after a catastrophic third world war that is referred to, within the strange Belial-worshipping society where the action takes place, as "the Thing." Animals and plants—and of course human beings—have been mutated, especially by gamma rays. Babies are ritually killed if their changes go beyond certain limits (such as having more than three pairs of nipples or more than seven fingers or toes).

Similar plot points appear in John Wyndham's *The Chrysalids*; however, this is a more conventional and realistic novel than *Ape and Essence*. The action takes place long after a nuclear war that is faintly remembered as "the Tribulation," and the story is told in the first person by David Strorm, who was brought up within a farming community in Labrador. This has relatively basic technology, even by the standards of the 1950s, with horses and carts, traditional kinds of animal breeding, and slow-loading guns. In many situations, the locals prefer bows and arrows to their unsophisticated firearms, since their bows can be reloaded more quickly despite having a shorter range.

The community is obsessed with maintaining the purity of form of its livestock, crops, and people: anyone who shows even the slightest deviation from the norm (as with the six toes of David's friend Sophie Wender) is driven out. While it is obvious to readers that mutation has been accelerated by widespread radioactivity, this is unknown to most of the characters in the story. Biological purity is seen as the basis of salvation, and the community reinforces this with slogans such as "WATCH THOU FOR THE MUTANT!" Anyone or anything categorized as a mutant is considered accursed.

We learn that children are taught an Ethics class, although it is nothing like ethics as twenty-first century philosophers understand it (for example, it bears no resemblance to my explanation in Chapter 3 of this book). Its gist is that humankind is obliged to follow a faint, difficult, and singular trail back to God's grace and the restoration of a Golden Age, after the penance inflicted by God in the Tribulation (again, readers, but not the characters, can interpret this as a nuclear war). The religious and secular authorities are empowered to judge each step on the trail to determine whether it is sinful or part of the true re-ascent.

David finds himself unable, in all seriousness, to classify his friend Sophie as a Deviation, and he helps her and her mother keep her mutation secret. As we can see from outside the narrative, the local system of moral requirements is based on unlikely theological and metaphysical premises. It has little to do with ordinary human sympathies or with what is really needed for the functioning of the community. Within the narrative, there is a difference between those characters who adhere rigorously and mercilessly to the local moral and theological requirements and those who are driven by human understanding and compassion. Although the latter approach does not pre-dominate, it is surprisingly common. Many people show some doubts about what they've been taught, and the doubts go along with decent impulses to help others. Even some of the harsher characters find it difficult to abide by the rules completely.

Although David is outwardly normal—as his society understands normal-ity—it becomes clear that he and various others, including his half cousin Rosalind Morton, possess a telepathic ability that enables them to connect over long distances. David's baby sister, Petra, possesses the same ability but much more strongly than others; she can communicate as far as "Sealand" (New Zealand) many thousands of miles away. The main story is about how their powers gradually become known to the community ⋯ and the community's savage response.

Mark Hillegas describes this novel's central idea as follows: "man as he now exists is an inadequate species which must be superseded if the world is to survive" (1967, 172). But this is too literal. The central idea, rather, is a plea for toleration, backed up by skepticism as to whether anyone really knows the mind of God. *The Chrysalids* is not a manifesto of atheism, but it is (among other things) a critique of theologically grounded morality.

The story ends on a hopeful note, but arguably with some foreboding. The Sealanders rescue David and some of his companions (for complicated reasons the rescue mission is left incomplete), and we are offered hope that the Sealanders have developed a better society than the one David has left behind. They enjoy telepathic communication among themselves, and this does seem to conduce to a kinder, more cooperative and creative way of life. And yet, the Sealanders are not presented as perfect. They show a high opinion of themselves, and they especially view themselves as superior to what they consider the Old People, with their individual isolation and their failure to cooperate effectively on a large scale.

The woman from Sealand who rescues David is clinical and pitiless about killing his enemies, and she sounds almost like a comic-book villain in proclaiming the merits of evolutionary change, the supersession of each life form over time, and the superiority of her own human variant.[3] Perhaps we're meant to take this at face value: Wyndham may be endorsing the sort of ethic of destiny that so bothered C.S. Lewis. Or perhaps not. Whatever Wyndham intended, there's a hint, as *The Chrysalids* come to its end, that all societies rationalize dubious values in sanctimonious ways, even while proclaiming their own righteousness.

Concluding Remarks

Science fiction's mega-text is a rich and thematically fruitful accumulation of icons and tropes. As Broderick explains, these do not have fixed meanings established by literary conventions. For example, robots have been presented in multiple ways. Nonetheless, an icon such as the robot seems to attract a relatively small range of possible meanings. Varied ways of imagining robots—or any other such icon—"tend to cohere about a limited range of narrative vectors" (Broderick 1995, 60).

Aliens in science fiction stories are often invaders, but sometimes they are our friends, sometimes we invade their territories, and sometimes they are just plain mysterious. These categories are not exhaustive, but stories about aliens tend to take a relatively small number of general forms and to engage recurrently with certain ideas. Imaginary aliens also show some common variations of appearance. As I mentioned earlier, the nastier ones are often reptilian or insectoid. But this is hardly compulsory. The alien mind in Lem's *Solaris* is an

---

[3] Wells's Marcus Karenin, in *The World Set Free* (2014), is an earlier version of the "good" character who speaks in this visionary way. For more, see Chapter 7.

incomprehensible ocean, and Gregory Benford has often depicted intelligent beings based on magnetic fields. In *The Sunborn*, interaction between such beings and human explorers leads the humans to speculate about what sort of moral system such radically different intelligences would develop.

The classic robot story involves rebellion, and the classic story of mutants involves their rejection by a world that fears them. However, authors and readers need not be stuck with those tropes. Often there is a counternarrative of integration, even if it requires someone fleeing to a more advanced society as in *The Chrysalids*. Science fiction interrogates itself—which is, perhaps, an overly oracular way of saying that science fiction writers respond to each other's storylines and ideas. Furthermore, however frightening aliens, robots, and mutants might seem to the characters confronted by them, they are usually enthralling from the perspective of the reader. This creates a pressure to write more positive stories about them. The result can be an implicit ethic that extends concern beyond human beings to all Intelligent Others.

References

Broderick, D. (1995). *Reading by starlight: Postmodern science fiction.* New York: Routledge.

Gomel, E. (2014). *Science fiction, alien encounters, and the ethics of posthumanism.* Basingstoke, Hampshire: Palgrave Macmillan.

Graham, E. L. (2002). *Representations of the post/human: Monsters, aliens and others in popular culture.* New Brunswick, NJ: Rutgers University Press.

Grech, V. (2014). The Pinocchio syndrome and the prosthetic impulse. In R. Blackford & D. Broderick (Eds.), *Intelligence unbound: The future of uploaded and machine minds* (pp. 263–278). Chichester, West Sussex: Wiley-Blackwell.

Hillegas, M. R. (1967). *The future as nightmare: H.G. Wells and the anti-utopians.* New York: Oxford University Press.

Jackson, P. T., & Heilman, J. (2008). Outside context problems: Liberalism and the other in the work of Iain M. Banks. In D. M. Hassler & C. Wilcox (Eds.), *New boundaries in political science fiction* (pp. 235–258). Columbia, SC: University of South Carolina Press.

Joy, B. (2000, April). Why the future doesn't need us. *Wired.* Accessed May 7, 2017, from https://www.wired.com/2000/04/joy-2/

Nichols, R., Smith, N. D., & Miller, F. (2009). *Philosophy through science fiction: A coursebook with readings.* New York: Routledge.

Williams, B. (1985). *Ethics and the limits of philosophy.* London: Fontana.

7

Going Inward: Science Fiction and Human Enhancement

Change and the Human Future

In Chapter 1, I drew attention to some inferences that might be drawn from the scientific and technological revolutions of recent centuries. It is now established that we inhabit an incomprehensibly vast universe whose origins lie deep in time. We are, ourselves, the product of natural events taking place over many millions of years. In all meaningful ways, we are continuous with other animal species, rendering pre-scientific ideas of human exceptionalism untenable. Furthermore, our particular societies and cultures are mutable. They have changed in the past—for reasons that now include intra-generational technological change—and that will continue.

While this set of claims might seem almost commonsensical to most readers, it is not pre-scientific common sense. It represents a dramatic historical shift in human understanding of the universe and our place in it. Not that long ago, such ideas would have been viewed in the Christian West (and most other parts of the world) as outrageously radical and intolerably heretical—and they still meet with resistance from many quarters. During the twentieth century, C.S. Lewis's *Out of the Silent Planet* and its sequels formed a small, but significant, part of that resistance. Lewis sought to portray an older image of the universe and humanity as still intellectually tenable, as well as emotionally attractive. More commonly, however, science fiction stories take place within a comprehensively revised picture of the world and our place in it.

The revised world picture suggests an additional insight that is currently controversial: the idea that human beings, as individuals or as a species, are

© Springer International Publishing AG 2017
R. Blackford, *Science Fiction and the Moral Imagination*, Science and Fiction,
DOI 10.1007/978-3-319-61685-8_7

themselves technologically alterable. The prospect of transforming ourselves and enhancing our capacities increasingly seems realistic. Research in such fields as genetic engineering, cognitive science, and psychopharmacology offers seemingly unlimited prospects. At some point, we may be able to make extensive modifications to human DNA, body tissues, or neurophysiological functioning, or to merge our bodies with sophisticated cybernetic devices ("cyborgizing" ourselves). Why might we want to do this? The most obvious goal is that of increasing the maximum human life span. More generally, we might want to enhance our athletic, perceptual, or cognitive abilities—or those of our children and descendants—to a point beyond any historical level. Depending on how the new technologies are used, they might alter the fundamentals of human nature and the human condition. All of this would likely happen in iterations—and thus be a new process of evolution—but on a timescale many orders of magnitude shorter than evolutionary change via natural selection.

But for all that I've stated so far, attempts to transform human bodies beyond a certain point might prove disastrous for the individuals involved and/or dystopian in their larger social impact. Let's consider how this plays out in science fiction.

Rejecting the Superhuman

The rise of powerful new technologies in the eighteenth and nineteenth centuries produced a mix of welcoming and cautious—even hostile— responses, and these were reflected in the new genre of science fiction. By the end of the nineteenth century, some writers were pondering the use of technology to change *ourselves*, not just as a tool to act on the world. One precursor was Nathaniel Hawthorne's short story "Rappaccini's Daughter" (1844), in which a scientist's daughter becomes poisonous to others by growing up in proximity to her father's poisonous plants. This, however, is not a deliberate attempt to alter a human being through technology.

Closer to my theme is Robert Louis Stevenson's short novel *Strange Case of Dr Jekyll and Mr Hyde* (1886). Here, the restrained and benevolent Dr. Henry Jekyll develops a chemical means of transforming himself into a man who purely embodies the animalistic and "evil" side of his personality. The resulting creature, Mr. Hyde, is small and subtly deformed, but also ferocious and strong. Since he does not exist separately from his creator, he can easily avoid detection for his crimes. In this respect, he resembles Griffin, the violent and dangerous individual at the center of H.G. Wells's *The Invisible Man*.

Jekyll's invention enables him to act with impunity on his worst urges (presumably sexual, although it's all kept rather discreet), even while distancing himself from them. Ultimately—who would have thought?—things go very wrong. In a rage, Hyde beats to death a distinguished politician, Sir Danvers Carew. The crime cannot be attributed to Jekyll, but he responds by ceasing his transformations. Soon, however, he finds himself transforming involuntarily, losing control of what he had set in motion. Unlike Victor Frankenstein, then, Jekyll does not create a creature physically separate from himself, but he is no less a Frankenscientist for that. Kind and gentlemanly though he is, Jekyll is a rogue figure.

Stevenson's main interest was in exploring ideas about psychology. However, *Strange Case of Dr Jekyll and Mr Hyde* was informed by the more general scientific ferment of the time, triggered—in large part—by the publication in 1859 of Charles Darwin's *On the Origin of Species*. Hyde shows considerable cunning when needed. In essence, however, he is a degenerate figure, a nineteenth-century conception of a lower, uncivilizable form of humanity.

Jekyll and Griffin are not very ambitious: they transform themselves into super-effective, though ultimately unsuccessful, criminals. Something on a larger scale—in more ways than one—is involved in Wells's *The Food of the Gods and How it came to Earth* (1904). The premise here is that two scientists invent a new food that they name "Herakleophorbia," although it becomes popularly known as "Boomfood" and the narrator refers to it as "the Food of the Gods." Herakleophorbia greatly accelerates the growth of any organisms that consume it. Events start to get out of hand when one scientist uses it to experiment on chickens. The married couple that he assigns to look after the chickens, and to administer their food, don't take the experiment seriously. They allow the substance to spread into the local environment, leading, for starters, to gigantic and menacing rats.

Some people, including Redwood, one of Herakleophorbia's scientific discoverers, experiment with the food as a dietary supplement to help produce strong babies, and the result, after two decades, is that the babies grow to enormous size—about forty feet tall. England and some other countries end up with populations of young giants, perhaps a few hundred in all, in the midst of the ordinary human beings. Meanwhile, local plant life is contaminated with Herakleophorbia, creating monstrous grasses, and there are ongoing problems with huge critters such as deadly wasps.

The narrator of *The Food of the Gods* has a distinctive voice: satirical, jocular, mocking of scientists, and skeptical about the benefits of science. The first half, or more, of the novel is straightforwardly funny. Wells exploits various events and situations for comic effect. As the narrative progresses, however, the tone changes. The narrator gradually begins to employ more sensitive and serious language, and the events themselves take a tragic turn. As they grow to adulthood, the young giants harbor no ill will toward ordinary humans and wish for nothing more than acceptance in the same way as other children and adolescents. However, the wider population comes to fear and resent them. Ordinary-sized humans take steps to register each of the giants, restrict their movements, and generally control their lives.

Inevitably, as it seems, this leads to hostilities. Though they are not vicious by nature, the giants chafe at their restrictions. One of them, Albert Edward Caddles, rebels against his menial work in a chalk pit. He sets off to explore the larger world, meaning no harm but no longer willing to be reined in. He is hounded and attacked. Eventually he defends himself and is shot dead by the police. Before he dies, Caddles repeatedly protests that he only wants to be left alone by the little humans, as he perceives them. (Here, he is a forerunner to a well-known Marvel Comics character, the Incredible Hulk, created in the 1960s. The Hulk frequently protests in the same way when he is hounded by human foes.)

Another sub-plot involves a love affair between two giants: one of them Redwood's gargantuan son, and one a visiting princess from continental Europe. By this stage, the narrator has settled into a lyrical style, and the two lovers are presented sympathetically. They are hounded, of course, and events come to a head after the demagogic Caterham is elected to power with a policy of hostility to the giants. His government demands that the giants live in isolation at a remote place to be determined, and that they have no children so their kind will die out after a generation. Giants and ordinary humans are soon caught up in a war whose conclusion is not shown, though the rhetoric of the final pages suggests that the giants will prevail.

The Food of the Gods can be read as a cautionary novel, showing how beings with truly superhuman power would necessarily come into conflict with the rest of us—not out of hostile feeling on the part of the superhumans, but because we would refuse to accept them. The satirical and comic tone of the novel's early chapters suggests that inventing Herakleophorbia was an act of foolishness. By the end, however, the emphasis has changed. Once the narrator begins to focus on events more from the viewpoint of the giants, they are revealed as tragic, unjustly treated individuals. With their forbearing attitude to the smaller people around them, they seem morally as well as physically superior to ordinary-sized humanity.

Thus, *The Food of the Gods* is a prototype of the narrative in which superior beings must cope with a world that hates and fears them. While Victor Frankenstein recoils at producing a race of monsters to compete with humanity, the new race of giants is unstoppable because the formula for Herakleophorbia has been widely disseminated and used. Even if the first generation of giants is destroyed, there will be more. This, however, is seen as a step forward in humanity's evolution.

By contrast, Philip Wylie's *Gladiator* is unambiguously pessimistic. It is the fictional biography of a solitary superhuman, Hugo Danner. Danner's father is a professor of biology at a small college in Colorado, and an expert in the physiology of insects. He postulates that modifications to human tissues could produce superhumans with similar proportional strength to ants or the leaping ability of grasshoppers. First, he experiments on a cat, which he then has to poison when it becomes a menace to the neighborhood. Undeterred, he experiments on his son before he is even born, with the result that baby Hugo demonstrates astonishing strength.

As Hugo Danner grows up and seeks his place in the world, his strength becomes a curse rather than a blessing. It is so out of scale with the strength of even the strongest men that any manifestations of it shock and frighten others. He can leap immense distances, bend steel objects, tear open a massive bank vault, and survive direct hits from bullets (which don't even penetrate his skin). When he enlists in World War I, first with the French Foreign Legion and then with the US army, he endures even the most extreme attacks from military weapons, receiving only superficial wounds.

Life in the trenches soon disabuses him of any romance about resisting Germany's imperial ambitions, and he comes to hate the carnage of war. With his strength and near-invulnerability, however, he is devastatingly effective as a soldier. War is the one area of human life where he can excel with little in the way of questions asked—much as Griffin's invisibility in *The Invisible Man* equipped him only for stealth and violence. When he returns to civilian life, Danner soon finds himself in trouble, much as he'd experienced at school and in college. Each attempt to use his great power to help others brings him grief.

Danner sometimes loses his temper in frustration, but he is otherwise well meaning. He finds himself, however, among people who hate and fear him (indeed, hate him *because* they fear him) as soon as they see any indication of what he can really do. Near the end of the novel, he makes one friend who accepts him for what he is—the great archeologist Dan Hardin. Hardin suggests a solution to Danner's plight not unlike that devised by Franken-stein's monster and perhaps found by Wells's giants in *The Food of the Gods*. That is, he should create a race of hundreds of his own kind: superhuman men and women operating within human society and giving each other support. He might, Hardin argues, be the beginning of a new breed of titans who could create their own society and culture, gradually coming to supersede ordinary humans and to dominate the world.

Though initially attracted by Hardin's vision, Danner is soon repelled by it, foreseeing hatred from normal humans, inevitable warfare between humans and titans, and perhaps catastrophic warfare among the titans themselves if they jostle for power in a new political order. In this mood, distraught and emotionally torn, he seeks isolation and calls out to God for guidance, leading immediately to his death by a lightning bolt.

This *deus ex machina* ending is unsatisfactory, spoiling what is otherwise an entertaining and thought-provoking novel. It is often claimed that *Gladiator* was the inspiration for the popular comic-book character Superman, created in 1938—eight years after *Gladiator*'s publication. However, this point has never been clearly established, and the creators of Superman doubtless drew on other influences whether or not they were influenced by Wylie.[1] Be that as it may, *Gladiator* was popular in its time and undoubtedly exerted direct and indirect influence on later science fiction. It clearly influenced Robert A. Heinlein when he came to write the story of another superhuman being who is ultimately destroyed: Valentine Michael Smith of *Stranger in a Strange Land*. Both characters spend time earning their livings through carnival acts, Mike as a magician and Danner as a strong man in a Coney Island sideshow.[2]

Like *The Food of the Gods*, *Gladiator* suggests that human beings can never accept the presence of other humans with personal abilities greatly beyond their own. With its repeated references to Danner's life among humans who hate and fear him because of his immense strength, *Gladiator* is also a prototype for the popular Marvel Comics characters, the X-Men. The X-Men are heroes whose stated mission is protecting a world that hates and fears superhuman mutants like themselves. As we saw in Chapter 6, the classic story involving mutants is their rejection by human society. This appears in, among other prominent texts, Olaf Stapledon's *Odd John*. We can now generalize further: many science fiction writers expect scenes of rejection and conflict if anyone ever acquires off-the-scale personal abilities, whether through random mutation or (as appears more scientifically plausible) some form of deliberate technological intervention.

Stapledon portrays the hatred and fear of human beings even for a technologically enhanced dog. Almost a decade after *Odd John*, he published *Sirius*, which tells the story of a highly intelligent canine who can hold his own in

---

[1] Another obvious (possible) influence is the work of Edgar Rice Burroughs.
[2] Interestingly, *Gladiator* includes a minor character called "Valentine Mitchel," but I'm unaware whether Heinlein named his hero partly as a nod to Wylie.

conversation with human scientists. This is a fascinating novel that merits more discussion than I can provide here. Suffice it to say that Sirius is the product of experiments by the great physiologist Thomas Trelone, a rather mild-mannered Galilean scientist when compared to Victor Frankenstein or the frightening Doctor Moreau—but an effective one. Like Frankenstein's monster, Odd John, and Hugo Danner, Sirius finds himself rejected by human society. The one note of optimism is that before the end he enjoys some rich and happy interactions with particular human beings.

By the middle decades of the twentieth century, then, one established theme in science fiction was that truly great personal abilities, beyond the human scale, are more a curse than a blessing. The inevitable result, some of these stories suggested, was rejection and/or violent conflict. At best—seen from the viewpoint of those with the great abilities—conflict might lead to victory over the rest of us. During this same period, however, a counter-narrative was taking shape.

Human Destiny and the Future of Mind

When we read his novels and stories, H.G. Wells does not always come across as a "Wellsian" techno-utopian, but one element in his thought was a vision of humanity's boundless prospects. In *The World Set Free*, published in 1914, he depicts a nuclear war—though with atomic bombs somewhat different from those invented three decades later—and the devastation of major cities across the world. Yet, as its title suggests, *The World Set Free* is ultimately an optimistic novel. The destruction of the old political order clears the way for a new one that includes a world state. Like much other science fiction, *The World Set Free* engages with large questions about the responsible use of science and technology; it suggests that we must place technological power in the hands of a world state or it will destroy us. Properly used, however, technology can lead to a self-directed evolution of mankind and a limitless destiny.

In its final chapter, the novel looks further forward, as it focuses attention on the visionary Marcus Karenin. Karenin imagines an astonishing future for humanity: he speaks of coming technologies that will grant us control over our bodies, abilities, and emotional responses, removing inherited weaknesses, including susceptibility to age, disease, ill-will, fatigue, and death. The poet Kahn protests, anxiously, that such developments would take us beyond humanity, but Karenin won't have it. He launches into a peroration about space travel that could easily have been placed in the mouth of a villainous mad scientist, except there is no indication that Karenin is anything other than deeply wise and perceptive (although he also seems rather crotchety and impatient). His conception of mankind's future, one in which we will enhance ourselves technologically and colonize the stars, is framed as entirely positive and inspirational.

By the 1920s, there was a strong intellectual movement in support of human enhancement through technology. In his manifesto *The World, the Flesh and the Devil*, first published in 1929, J.D. Bernal wrote persuasively of our past use of technological objects to modify what he called "an effective human body." According to Bernal, humanity's ways of acting upon the world, experiencing it, and perceiving it have all been modified by innovations such as stone tools, clothing, and (much more recently) spectacles. He observes, however, that all of these lie outside the body's living cell layers. Accordingly, "The decisive step will come when we extend the foreign body into the actual structure of living matter" (Bernal 1970 [1929], 33). He goes on to imagine a more capable human body. This, he claims, is required as a more efficient way to sustain our mental activity:

> the increasing complexity of man's existence, particularly the mental capacity required to deal with its mechanical and physical complications, gives rise to the need for a much more complex sensory and motor organization, and even more fundamentally for a better organized cerebral mechanism. Sooner or later the useless parts of the body must be given more modern functions or dispensed with altogether, and in their place we must incorporate in the effective body the mechanisms of the new functions. (Bernal 1970 [1929], 35)

In *The World, the Flesh, and the Devil*, Bernal frequently cites the contemporary writings of his fellow scientist J.B.S. Haldane, especially the latter's vision of the human of the future: born, and originally nourished, in an ectogenetic[3] factory. Bernal imagines bodies thoroughly mechanized and transformed by technology, and indefinitely open to further modification. These might take a range—perhaps a wide range—of specialized forms. Even beyond this, individual brains might be brought into direct thought-to-thought contact via technological connections. Bernal even imagines various paths by which humanity might split into two or more species. Humanizers and mechanizers might take separate paths. If so, conflict between the rival groups "will be solved not by the victory of one or the other, but by the

---

[3] I.e., developing as embryos in an artificial environment outside the womb.

splitting of the human—the one section developing a fully balanced humanity, the other groping unsteadily beyond it" (Bernal 1970 [1929], 56).

This is just one version among the many science fiction narratives that imagine a split within humanity. Such narratives go back at least as far as *The Time Machine* and arguably even back to *Frankenstein*. The most obvious difference between Bernal and Wells (in *The Time Machine*, though certainly not in *The Food of the Gods*) is that Bernal imagines a deliberate process of technological intervention. By contrast, Wells's Morlocks and Eloi differentiate into separate species through an unforeseen and unwished molding by evolutionary forces. The Morlocks emerge gradually from industrial workers increasingly working below the ground, while the graceful, beautiful, but helpless Eloi emerge from pampered surface dwellers. While the Eloi began as the masters, in Wells's scenario they ultimately become the Morlocks' cattle. Yet both species devolved, in dramatically different ways, from present-day *Homo sapiens*.

Like Wells's Karenin, Bernal contemplates that an emergent, more scientifically oriented and advanced, species might leave Earth for the stars. This idea exerted enormous influence on the subsequent development of science fiction.

The literary critic George Slusser has written a long and perceptive article on science fiction's visions of human transformation: "Dimorphs and Doubles: J.D. Bernal's 'Two Cultures' and the Transhuman Promise" (2009). Slusser examines works by a variety of authors, and argues that Bernal sketched a master plot for the subsequent development of the genre. Thereafter, science fiction commonly depicts a "dimorphic split" in the human species: at a crucial moment in history, forward-looking individuals evolve rapidly into something other than human, while others stay behind, retreating from the promise of transformation. This lends itself to endless variations, subversions, and complications.

We can, for example, see elements of this plot throughout Arthur C. Clarke's fiction and nonfiction. Clarke was a technological meliorist, a painstaking engineer of utopian dreams who showed a clear understanding of well-established science. In this role, he advocated the exploration and colonization of space, and he expressed hopes for a coming age of material abundance. At the same time, his work includes a more obviously visionary element. He can be seen straining at the limits of the possible and speculating about the transformation or supersession of humanity. In this mood, Clarke imagined humanity's eventual transformation into something new and strange—or, alternatively, our replacement by non-human intelligence.

In his essay "The Obsolescence of Man," included in *Profiles of the Future*, Clarke writes that "Biological evolution has given way to a far more rapid process—technological evolution. To put it bluntly and brutally, the machine is going to take over" (2000 [1962], 194). He does not predict an active rebellion by robots, as depicted in so many novels and movies, but he does speculate that humanity is approaching its supersession by machine intelligence. Alternatively, we will have to transform ourselves into something far more capable than we are now, with our current bodies. Clarke's vision—and with it his implicit ethic of destiny—is captured succinctly in another passage from *Profiles of the Future*, where he acknowledges a criticism by the historian Lewis Mumford of the prospect of space colonization. Clarke responds memorably:

> But when [Mumford] wrote: "No one can pretend ⋯ that existence on a space satellite or on the barren face of the Moon would bear any resemblance to human life," he may well be expressing a truth he had not intended. "Existence on dry land," the more conservative fish may have said to their amphibious relatives, a billion years ago, "will bear no resemblance to piscatorial life. We will stay where we are."
> They did. They are still fish. (2000 [1962], 90.)

For Clarke, the colonization of space will encourage us to become something more than human, as fish evolved into forms that were more than fish.

Much of Clarke's work contains radical speculations about the ultimate future of life and intelligence, including the seemingly mystical suggestion that mind might exist in forms that do not depend upon matter as we know it. He portrays disembodied mentalities in some of his greatest novels: *Against the Fall of Night* (1953; expanded from a shorter version published in 1948); its lengthier, rewritten version, *The City and the Stars* (1956); and *Childhood's End* (1953).

Even in *Childhood's End*, where mind is shown as surviving in a "paraphysical" form, Clarke painstakingly distances himself from the traditions of mysticism. He did, however, take the idea of the paranormal seriously at some stages of his career. Like many other science fiction writers of the Campbell Golden Age and the 1950s, he introduced mental powers, such as telepathy, in much of his fiction, speculating that they might have some scientific rationale, as yet unknown. This tendency still appears in much science fiction, even of sorts that are otherwise distant from fantasy or any kind of occultism. On that basis, Clarke sometimes depicts the ultimate form of intelligence as lying outside physical, but not necessarily systematic or scientific, explanation.

In *Against the Fall of Night* and *The City and the Stars*, he portrays a future world separated from ours by a billion-year abyss of time. In *Against the Fall of Night*, Alvin, the main character, is the first child to have been born in the city of Diaspar for thousands of years. The city's inhabitants are immortal, and the corollary of this is that they do not reproduce. In *The City and the Stars*, Alvin's uniqueness is presented in a more complex and imaginative scenario. What makes him a "Unique" within the city is that his personal characteristics have been created for the first time, rather than recreated from a previous human template by the city's matter organizers, which use information on the billions of humans stored in Diaspar's memory units. Thus, even in the 1950s, Clarke was imagining stored human personalities in a computer substrate, finding inspiration in what were then new ideas in the field of information theory (see Regis 1990, 150).

In *Against the Fall of Night*, Clarke shows a powerful immaterial mind, Vanamonde, as the product of advanced scientific experiments far in Earth's future. However, he does not otherwise attempt to rationalize the creation of immaterial minds. In his 1991 authorized sequel, *Beyond the Fall of Night*, Gregory Benford presents the immaterial minds in the book as beings composed of complex magnetic fields. But Clarke takes a different tack in *The City and the Stars*. Here, Vanamonde is shown to be the product of manipulations of the structure of space itself in order to imprint the existence of mind upon a substrate that is more basic than matter. This is of a piece with Clarke's subsequent fiction and nonfiction writings in which he speculates that information and mind might be able to exist in forms that lie within the boundaries of physics but do not depend on matter as we know it.

In the Space Odyssey series—commencing with the movie (dir. Stanley Kubrick) and novelization of *2001: A Space Odyssey* in 1968—Clarke describes the evolutionary stages of alien intelligences that shaped humanity's own evolution. They have advanced from biological life, to a cyborg form with biological brains in machine bodies, then to a form of machine intelligence— ultimately recreating themselves as energy. For Clarke, this might also be the long-term human destiny.

Imagining Immortality

In myth, legend, and literary narratives, the quest for immortality is often portrayed as hubristic, and attainment of immortality is typically shown as a curse. Well-known examples include Tithonus in Greek mythology and the

Struldbruggs in Jonathan Swift's *Gulliver's Travels*. In each case, immortality without eternal youth is shown as a horrible fate. Aldous Huxley's *After Many a Summer* (1939; also known as *After Many a Summer Dies the Swan*) updates the idea and portrays the quest for immortality as a delusion. The dysfunctional cast of characters includes an especially unpleasant Galilean scientist and his wealthy patron, as well as the latter's young mistress. When they finally track down an immortal man, he has transformed over two centuries into something inarticulate and monstrously simian.

Overall, the Western cultural tradition portrays immortality on Earth as something forbidden to mortals. Immortality can be obtained, if at all, only in an afterlife. One high-literary exception is Virginia Woolf's treatment of her immortal protagonist in *Orlando* (1928), but this is a novelistic *jeu d'esprit* whose thematic intentions are remote from any serious comment on immortality.

George Bernard Shaw's five-play sequence *Back to Methuselah* takes a stand in favor of immortality. This complex work was influenced by Wells's writing, and it makes explicit reference to *The Food of the Gods*. It depicts past, present, and future stages in humanity's evolution. The past is represented by scenes involving Adam, Eve, Cain, and the Serpent of Eden—establishing a mythic and allegorical framework that influences even the action set in the author's present (the years immediately after World War I). While Shaw's long preface is scathing about strictly Darwinian theory, the entire sequence of short plays is grounded in nineteenth- and early twentieth-century debates about the mechanisms driving evolution. Shaw acknowledges a role for natural selection, but he favors a broadly Lamarckian view in which evolutionary change happens swiftly whenever it is needed and willed.

The preface explains *Back to Methuselah* as a first attempt to produce a myth—for Shaw, a culturally resonant narrative that is no worse for not being literally true—in the service of a new religion of Creative Evolution. Here, Shaw follows the thinking of French philosopher Henri Bergson, who imagined a vital force—an *élan vital*—driving evolution forward.

Though Shaw describes Creative Evolution rather vaguely in his preface, he gives it more concrete form in the action and dialogue of the plays. His characters are types, rather than fully rounded personalities, and the action of *Back to Methuselah* is somewhat schematic. Nonetheless, it shows what successive stages of human evolution might be like. In the present, the characters argue over a claim that human beings must obtain longer lives—ideally three hundred years—to obtain enough wisdom to organize complex modern societies. In the final three plays, set at progressively more distant future dates, we are shown a human destiny of continual striving for improvement. By 31,920, a truly strange society has developed, one in which human beings come to full maturity in only four years then live enormously long lives. These people are, however, seeking even greater longevity, and with it greater capability and wisdom.

The dissatisfaction expressed even by the extraordinarily wise and powerful elders of the year 31,920 might suggest that humanity's striving is futile. That, however, is not Shaw's final message. On the contrary, he depicts perpetual dissatisfaction and striving as a source of progress. Dissatisfaction and striving will lead to humanity's ultimate liberation in a form that transcends ordinary matter. As the final play in the sequence comes to an end, it is suggested that our distant descendants will become intelligent patterns of energy free to explore the universe.

While the ideas of Creative Evolution and an *élan vital* are scientifically discredited, *Back to Methuselah* exerted substantial influence, mainly indirectly, on the course of twentieth-century and twenty-first century science fiction. In reality, there is no prospect that human beings can gain impressively greater longevity and capability merely by acknowledging the need (assuming there is one) and willing it to happen. There is, however, a possibility that similar results, amounting to a form of self-directed evolution, might be obtained through technological means such as genetic manipulation. Thus, Shaw's ideas about human destiny need not stand or fall with Bergson's philosophical speculations.

It is not clear how many contemporary science fiction writers are directly familiar with Shaw's play sequence, but it almost certainly influenced Bernal and Stapledon (particularly the latter's *Last and First Men* and its sequels *Last Men in London* (1932) and *Star Maker* (1937)). *Last and First Men* depicts a future history of humanity that covers a total of eighteen human species. This goes far beyond what is shown in *Back to Methuselah*, with its three progressively more distant futures, but it is a logical extension of the same idea. Stapledon was, himself, influential on the SF writers who followed, both directly and indirectly via prominent authors such as Clarke. In fact, Clarke's vision of the ultimate destiny of human intelligence seems indistinguishable from Shaw's.

By contrast, *The Makropulos Affair*, a 1922 play by the Czech writer Karel Čapek, projects a negative view of extreme life extension. In this case, a life of three hundred years was bestowed on one character in the distant past—through an alchemical formula rather than anything more reputably scientific. The formula has worked its purpose, but such a long life turns out to be a curse. It produces cynicism and emotional coldness. Another character, Vitek, sees extended longevity as a source of hope—it could allow individuals to live fuller lives, rather than being struck down when they need more time for their life's work—but his is a minority voice within the play. Other characters have more selfish views, wanting to destroy the formula, to save it for an elite, or to exploit it commercially. Vitek sounds, in fact, very like the idealist Domin in Čapek's *R.U.R.*, with his futile dream of using robots to free humankind from drudgery and want.

The Makropulos Affair was first produced about the same time as *Back to Methuselah*, which was published in 1921 but not produced on stage until 1922. Indeed, Čapek comments on Shaw's play sequence in his preface to *The Makropulos Affair*. Čapek also raises strikingly contemporary concerns about the disruptive social effects of greatly extended life if its secret ever became widely known. Between them, therefore, Shaw and Čapek established the poles for an ongoing debate implicit within science fiction: immortality might provide extraordinary opportunities, or it might be the source of ultimate boredom or worse.

Overall, the popular view—even within the science fiction field, with its openness to change—still seems to be that immortality would be a curse and that pursuing it on Earth is forbidden. However, many SF narratives point in the other direction, among them Heinlein's *Methuselah's Children*—with its vision of a greatly extended span of human life—and James Blish's Cities in Flight series of novels, beginning with *They Shall Have Stars* (1956). By now, an extensive list of texts could be ranged on each side of the debate. The theme of immortality in science fiction could sustain a separate book-length study.[4]

Reshaping the Human

Science fiction has long shown teratological impulses, and this continues. However, tales of dangerous monstrosities, such as *Frankenstein*, have been balanced by more optimistic works depicting the use of advanced technology to reshape human bodies. One example is Blish's stories of pantropy—humans adapting themselves to other worlds—beginning with his short story "Surface Tension" (1952) and novel (based on earlier stories) *The Seedling Stars* (1956). Along with Clifford Simak's "Desertion" (1944), this founded a classic SF

---

[4] Meanwhile, see the various contributions to Slusser, Westfahl, and Rabkin (eds.) *Immortal Engines* (1996).

trope of humans altering their form to suit other worlds, rather than transforming those worlds to suit them.[5] Other examples include Joan Slonczewski's *A Door into Ocean* and Alison Sinclair's *Blueheart* (1996).

A more unsettling variation on the theme is Frederik Pohl's *Man Plus* (1976). At a time of global tension and environmental collapse, the novel's protagonist, Roger Torraway, is subjected to a slow, painful cyborgization process to adapt him for survival in the Martian environment. Plans to colonize Mars offer one glimmer of hope for humanity's future. *Man Plus* is sometimes harrowing for the reader, but its attitude to Torraway's enhancement is not entirely negative—merely realistic and unsentimental. This is also, in its way, another story of machines taking over, but it is more in the mode of Asimov than Čapek: Earth's networked computers secretly manipulate events to ensure humanity's survival, and hence the survival of their own kind.

More commonly, science fiction depicts human bodies that are reshaped or augmented for superiority in combat. In Alfred Bester's *The Stars My Destination*, for example, the vengeful and anti-heroic Gully Foyle has his body electrically upgraded to the point where his movements appear as blurs, making him, in effect, a deadly killing machine. Similar ideas have been used for many other heroes and villains in science fiction and the superhero genre. *The Stars My Destination* was a strong influence on the cyberpunk movement of the 1980s, though by then enhancements such as Gully Foyle's were already old news to science fiction fans.

During the 1980s and 1990s, the biological and computational sciences were becoming especially prominent within the intellectual cultures of industrialized nations. This reflected scientific advances—including the rise of rapid, computerized communications—and it involved some foreboding about where biotech and computers might be leading us. These concerns and emphases were reflected in the science fiction of the time, and the trend has continued to the point where no one reader can keep up with all the variations. As with the theme of immortality, this could sustain a lengthy study in its own right. In the remainder of this section, I'll comment on just a few revealing examples.

Observation of the rapid pace of technological advance led to the idea, popularized by Vernor Vinge in particular, that humanity is headed—sooner rather than later—for a technological singularity (or "the Singularity"), a point of practically infinite technological possibility. This might come about if

---

[5] In fact, the idea turns up as early as Stapledon's *Last and First Men* in 1930, although this describes an attempt—ultimately not successful—by one successor species to humanity as we know it to engineer another such species that is capable of surviving on Neptune.

further progress falls under the control of minds increasingly more powerful than ours. Vinge's prolific speculations about such issues, initially in his novella "True Names" (1981) and his novel *Marooned in Realtime* (1986), made him something of an icon for pro-technology activists.

Damien Broderick's *The Judas Mandala* (1982; revised in an Australian edition in 1990) depends upon the elaborately worked out convolutions of a plot involving time travel and alternative time lines. It depicts the efforts of the human characters to prevent a future world ruled with spiritless paternalism by seemingly omnipotent cyborg lords. The main character, Maggie Roche, is a poet, a martial artist, a self-assertive feminist, and a self-described utopian anarchist. Maggie is clearly a rebel, though she is not a modern Prometheus in the tradition of Victor Frankenstein. On the contrary, she is more an anti-Frankenstein, attempting to channel development of technology in a more responsible and benign direction. The Frankenstein role is assigned to David Elfield, who, as a young scientist before Maggie was even born, set himself on a path to becoming the Daystar—a cold, godlike cyborg of the future.

Broderick's younger Australian countryman, Greg Egan, emerged in the 1990s as a key exponent of the hard science fiction style. One ever-present aspect of Egan's vision is his insistence that human identity and character are ultimately reducible to the functioning of our neurophysiology, beyond that to the past processes of biological evolution, and, even beyond those, to the fundamentals of purely physical entities, events, and causal influences. This idea is at the center of his fiction, and his characters often face the unsettling fact that their personalities and other aspects of their minds can be altered by physical interventions. In these narratives, technology is shown as more than a tool. It can act directly on human brains to alter personalities or assist cognition. This invites large questions about what it is to be human, what we most value in our humanness, and whether humanness—whatever that amounts to—is inherently valuable in any event.

One example can be found in *Quarantine* (1992), Egan's first novel to appear from a large international publisher. Here, he imagines the implications of neural modifications, or simply "mods," which can reconfigure an individual's personality. In the novel's backstory, its protagonist, Nick Stavrianos, faced a bizarre choice when his wife whom he deeply loved was killed while his mind was modified in a way that blocked him from caring. Though he was not emotionally hurt by her death, he knew that he would be as soon as he returned to his normal mental state. To protect himself from grief, he acquired another mod which he continues to use during the main action of the novel. The new mod was crafted to duplicate the happiness that his wife formerly gave him. As Stavrianos justifies his decision, mental states such as happiness and grief are just matters of neural anatomy. Why not control them by whatever means are possible?

Nancy Kress was another key figure in the 1990s renaissance of hard SF. Among her achievements is a body of fiction that commenced with her novella "Beggars in Spain" (1990). Subsequently, she expanded this into a successful novel, *Beggars in Spain* (1993), and then into two sequels, *Beggars and Choosers* (1994) and *Beggars Ride* (1996)—thus forming a trilogy. The result is a personal, yet explicitly political, study of what might happen if human abilities were enhanced by genetic engineering. The main characters are modified human beings whose superior abilities include the capacity to function without sleep and devote the extra hours to self-improvement. It also turns out that they do not age, and so they can live indefinitely. Kress imagines the "Sleepless," their creators, and their society—and of course, its response—in impressive detail that provides food for philosophical thought.

Thus, she engages with such questions as how human societies should—and in any event, *would*—react to the presence of individuals who have innate advantages over everyone else in whatever fields they choose to work. Conversely, what duties would such extraordinary individuals owe to those less fortunate? How far should they sacrifice the integrity of their own impressive lives in working to help others? All this suggests broader questions that arise even in a world without genetic engineering. What do highly capable and productive people who attract rewards for their efforts owe to people who are not socially productive as a result of their circumstances or lesser talents? Is it ethically okay for the talented simply to "move on" in their lives, focused entirely upon their own success and ignoring the needs of others?

Kress dramatizes the issues through vivid characters who are faced with high-stakes choices. Not least among the *dramatis personae* is Jennifer Sharifi, a leader of the Sleepless who plays the role of charismatic villain through much of the trilogy. Sharifi is willing to take extreme steps to protect her people, and these are clearly condemned by the logic of the narrative. However, Kress offers no optimistic alternative. As in many other works of science fiction—the prototypes include Wells's *The Food of the Gods*, Wylie's *Gladiator*, and Stapledon's *Odd John*—conflict appears inevitable once some people have personal abilities beyond the human scale. At the same time, much is lost if these people are destroyed.

At this point, we should also consider the idea of mind uploading, a relatively recent, but well-established, addition to the SF mega-text. Here, the claim is that a human personality and identity could be, as it were, transferred from the organic substrate of a brain and neural system to very advanced, powerful computer hardware. This could, supposedly, have advantages in terms of greater durability and longevity, accelerated speed of thought, and enhanced physical and perceptual capacities (assuming that the computer is attached to sensors and to some kind of body or set of tools that can act on the world). Much science fiction raises questions about whether such an uploaded personality would be conscious and whether (even if so) it could truly claim to be the same person who was originally uploaded. William Gibson's "The Winter Market" (1986) is a classic story dealing with the issue of continued identity. There is also an extensive body of philosophical literature debating these points.[6]

The 2014 movie *Transcendence* engages with similar issues. This rather cerebral work performed badly at the box office, but that is not a fair measure of its artistic merit or its entertainment value. *Transcendence* is an involving and intelligent film that deserved to attract a larger, more sympathetic audience. At first, the plot revolves around the efforts of Dr. Will Caster, yet another Galilean, and potentially Frankensteinian, scientist. Will is engaged in Artificial Intelligence research, attempting to develop a fully conscious AI to run on a powerful quantum computer. However, a dramatic series of events leads to a change of plans, and Will's wife, Evelyn, uploads his mind into the computer substrate. Then, at the request of the newly uploaded "Will," she connects his consciousness to the Internet.

Will—if, indeed, it is Will—quickly obtains great power from his ability to think at immensely accelerated speeds. He makes scientific and technological breakthroughs and attempts to establish a global utopia of plenty. However, he soon attracts suspicion and hostility. The problem for the other characters involved in the story is how anyone can know whether they really are dealing with Will, not merely a copy of his memories and personality or an AI based on them. More pointedly, are the uploaded Will's intentions toward humanity as benevolent as they first seem? There is soon an effort, with Evelyn's involvement, to circumvent Will's formidable abilities and shut him down. Human beings act against Will because, quite understandably, they fear him. It's a familiar story of rejection.

By contrast with such stories, post-cyberpunk space opera includes some optimistic visions of advanced societies where many kinds of intelligent beings live in harmony, and where problems with immortality or greatly extended longevity have been overcome. Iain M. Banks's Culture series portrays a vast interstellar civilization in which human beings who are enhanced to various

---

[6] For a comprehensive introduction to the debate, see the various contributions to Blackford and Broderick (eds.) 2014. My introductory chapter to this volume provides a succinct and up-to-date summary of the main arguments and what is at stake.

levels cooperate with Intelligent Others of various levels of capability, up to and including the (artificial) Minds that make the most important decisions for the Culture as a whole. Indefinitely long spans of life are an option, though most humans do choose to die after very long lives by our standards.

Banks's novels and stories are packed with conflict, but usually between the Culture's operatives and less enlightened civilizations—beginning, in *Consider Phlebas*, with the warlike and expansionist Idiran Empire. Despite its great diversity of intelligent beings, the Culture is internally peaceful. However, Banks offers little hint as to how our own planet could reach such a point without going through existential upheavals such as Kress and others depict. The Culture is, indeed, a bravely utopian vision, though in some ways Banks's approach is remarkably unsentimental. Few other science fiction writers are as unflinching in offering dark accounts of espionage and wars.

Concluding Remarks

Much of what we read in myth, legend, classic literary fiction, and even genre SF appears cautionary about human desires or attempts to exceed our "given" (in reality, biologically evolved) limits. If we attained immortality, it would be a curse—so the message seems to run—and likewise for enhancements such as truly superhuman strength. Furthermore, science fiction continually reminds us, great power does not entail great moral virtue: it might, on the contrary, lead to great social alienation. Many popular narratives insist that those with power should use it responsibility, but much in the SF canon suggests a more cynical lesson. Great power might loosen bonds of mutual concern or encourage ruthless actions by those who can get away with them. The old story of the Ring of Gyges comes to mind. This is, for example, a dismal conclusion that we might draw from reading Wells's *The Invisible Man*. If so, technologically enhanced humans might be our oppressors rather than benefactors.

Yet the canon of science fiction contains images of superheroic figures and daring or wise immortals. Not all science fiction writers withdraw in horror at the idea of thoroughly enhanced humans, made smarter, stronger, or longer-lived through applications of technology. Much SF portrays technologically altered human beings—possibly with very long lives—seeking humanity's destiny in space. Societies of the distant future might, for all we know, not be restricted to "baseline" humans. We can readily imagine societies populated by biologically enhanced humans of various kinds, along with ordinary human beings and a range of Intelligent Others including fully conscious Artificial Intelligences. This is, perhaps, a destiny worthy of dreaming.

But our ability to picture this in our minds—or to project it on a cinema screen—does not entail that, in the harsh light of reality, all these future people could prosper together side by side. Even if they could, how do we get to *there* from *here* except on a path through ruinous conflicts such as those described by H.G. Wells and Nancy Kress, among others? That kind of question can tease us out of thought.

References

Bernal, J. D. (1970). *The world, the flesh, and the devil: An inquiry into the future of the three enemies of the rational soul.* London: Jonathan Cape (Orig. pub. 1929).

Blackford, R., & Broderick, D. (Eds.). (2014). *Intelligence unbound: The future of uploaded and machine minds.* Chichester, West Sussex: Wiley-Blackwell.

Clarke, A. C. (2000). *Profiles of the future: An inquiry into the limits of the possible* (Millennium paperback edition). London: Indigo (Orig. pub. 1962).

Regis, E. (1990). *Great mambo chicken and the transhuman condition: Science slightly over the edge.* London: Viking.

Slusser, G. (2009). Dimorphs and doubles: J.D. Bernal's 'Two Cultures' and the transhuman promise. In G. Westfahl & G. Slusser (Eds.), *Science fiction and the two cultures: Essays on bridging the gap between the sciences and the humanities* (pp. 96–129). Jefferson, NC: McFarland.

Slusser, G., Westfahl, G., & Rabkin, E. S. (Eds.). (1996). *Immortal engines: Life extension and immortality in science fiction and fantasy.* Athens, GA, and London: University of Georgia Press.

8

Conclusion: Great Power and Great Responsibility

Broad New Canvas

The future is not certain, but human cultures and societies will certainly change. At a minimum, new technologies will alter how we live and how we experience the world—as happened with the invention of agriculture, writing, the printing press, steam engines, electrification, motor vehicles, radio and television, the contraceptive pill, digital computers, and social media. Debates about our self-conception as a species and about our place in the world will continue as science attempts to explain human beings and human nature in evolutionary, genetic, neurophysiological, and quasi-computational terms.

More controversially, there is now the prospect of altering ourselves or our children through deliberate, direct technological interventions. Until now, direct technological alterations of the human body have been trivial or, in a broad sense, therapeutic, though the line between therapy to restore health and enhancement to augment capacities is not clear,[1] and some widely accepted applications of medical technology already do more than restore health. Vaccination augments resistance to disease, rather than treating someone who is already sick, while the contraceptive pill is used (in its standard application) to control fertility, not to treat disease or injury. There are real possibilities for increasing use of technology to alter human bodies in ways that enhance cognitive and physical capacities.

---

[1] For an introduction to the therapy/enhancement distinction, and the difficulties that arise in attempting to draw a clear distinction, see Blackford 2014, 195–212.

© Springer International Publishing AG 2017
R. Blackford, *Science Fiction and the Moral Imagination*, Science and Fiction,
DOI 10.1007/978-3-319-61685-8_8

New understandings of the universe and ourselves, together with new technologies, have provoked new philosophies and cultural movements. Hence, we now see varieties of "posthumanism" and "transhumanism"—although it is often unclear how these terms are being used. The general idea of transhumanism is that we should *welcome* the use of technology to expand human capacities. In the extreme, emerging technologies might give us (or perhaps our children) abilities raised far beyond the current human range—and in that sense we (or our children) could meaningfully be considered "posthuman." If this is done properly, transhumanists argue, it will be a *good* thing. The suffix "trans" in "transhumanism" refers to a desired *transition* to a posthuman level of competence (Blackford 2013).

If transhumanism promotes a transition from *human* limitations to a higher level of physical and cognitive functioning, posthumanism is best understood as a critique of *humanism*. It is a set of philosophical positions that attempt to understand the world and guide our lives while repudiating the idea that human beings are ontologically exceptional. Posthumanist philosophical positions are, therefore, based on rejecting the anthropocentrism and human exceptionalism that posthumanists identify in the tradition of humanism.[2]

Posthumanism should not be confused with transhumanism. Indeed, the posthumanist scholar Cary Wolfe is hostile to transhumanism, which he labels "an *intensification* of humanism" (2010, xv). That said, repudiation of human exceptionalism is not logically inconsistent with welcoming technological efforts to enhance, for example, human longevity, intelligence, and athletic potential. It is logically possible, and perhaps also psychologically possible, to adopt a stance that is simultaneously posthumanist and transhumanist in the senses that I've briefly sketched.

To add to the terminological confusion, N. Katherine Hayles's well-received book *How We Became Posthuman* (1999) is a critique of attempts to understand the world and ourselves in terms of information theory. These attempts underpin some transhumanist ambitions, notably the idea of mind uploading, but Hayles has a wider agenda than merely critiqueing transhumanism and its goals.

My aim in this chapter is not to settle such terminological disputes or to argue for, or against, any particular philosophical position that might fall within posthumanism or transhumanism. My point is to highlight a revised

---

[2] But note, this is not the occasion to investigate whether posthumanist critiques of humanism are accurate and fair.

understanding of the human situation that has emerged through the period of European and Western modernity, especially under the influence of the natural sciences. This revised understanding makes posthumanist and transhumanist philosophies seem tenable and attractive, and it also underlies the attractions of science fiction. As far as I know, few science fiction writers identify explicitly as posthumanists or transhumanists, but many engage with posthumanist and transhumanist questions. They engage, that is, with the questions about the demise of human exceptionalism and with questions about the future of humanity as a species.

Science fiction writers were doing this before posthumanism and transhumanism were heard of. If anything, transhumanism was influenced by ideas in twentieth-century science fiction. Transhumanism emerged in something like its current form during the 1980s. By contrast, we can find proto-transhumanist ideas expressed by H.G. Wells early in the twentieth century and soon after by authors and thinkers such as George Bernard Shaw and J.D. Bernal. These ideas influenced Olaf Stapledon, Arthur C. Clarke, and many others involved in science fiction.[3]

Our enlarged understanding of the universe has given science fiction writers a very broad canvas. They can work with the vastness of space and the depths of time, and they need not assume current *mores*, levels of technology, or limitations on human abilities. It is natural for SF writers to speculate about life on other planets, contact between humans and alien life forms, and the possibility of human expansion beyond the confines of planet Earth. It is likewise natural for them to speculate about how humanity itself might change, whether from further biological evolution or from deliberate techno-logical interventions in the human body.

Science fiction's broad canvas enables writers to portray decisions of great moment, sometimes involving the future of humanity as a whole or of entire human or non-human civilizations. Stories of momentous choice can be cognitively valuable—they can challenge assumptions and provoke thought—but they can also lapse into mere spectacle. This raises difficult questions about the value of the SF genre and how we ought to engage with it critically.

---

[3] Note, however, that transhumanism and posthumanism have their own precursors prior to the twentieth century. A separate book would be needed to investigate and explain these.

The Sense of Wonder and Its Debasement

Science fiction's much-touted "sense of wonder" often involves awe at the immense scale of the non-human universe and the relative smallness of humanity. This feeling does not follow only from the sheer size of natural phenomena. It can, for example, be inspired by the breathtaking scale of futuristic or alien engineering or by the stakes involved in future conflicts.

Wells's *When the Sleeper Wakes* provides vivid descriptions of a future London, portrayed as a megacity with colossal buildings, vast chasms in between, and suspension bridges across the chasms. Wells provided a template for many such futuristic cities, including those shown on screen in *Metropolis* and *Blade Runner*. These can evoke emotions of awe and wonder. Similar emotions arise from descriptions—or visual portrayals—of huge spaceships and space stations, and similar gargantuan structures in space. Some science fiction narratives present engineering feats on an even grander scale. The classic example, but now one of many, is the eponymous megastructure in *Ringworld*, by Larry Niven, published in 1970. The Ringworld is an artificial ring with a star at its center and a diameter similar to that of Earth's orbit.

In some science fiction novels, entire planets are transformed to become more suitable for human life. This idea goes back at least to 1930, with the publication of Olaf Stapledon's *Last and First Men*. Another classic example is Arthur C. Clarke's novel *The Sands of Mars* (1951), which deals with events at a research base on Mars some decades in the future. It includes such innovations as turning one of Mars's moons, Phobos, into a second sun in order to warm the planet.[4] Karel Čapek's *War with the Newts* provides an early variation on the theme: millions of the salamander-like Newts are put to work under conditions of slavery, building huge landmasses in the ocean. But the Newts then seek to reconfigure the landmass of the Earth to suit the needs of *their* population—providing more area of shallow coastal water, since they cannot live on land or in the deep sea.

Enormously consequential choices have long figured in science fiction narratives, so much so that it is hardly necessary to give examples: many are described in earlier chapters. Rogue scientists such as Victor Frankenstein make choices that could put humanity itself at risk. Other heroes in science fiction act like Promethean benefactors to humanity, or perhaps as its existential destroyers. Thus, Gully Foyle in Alfred Bester's *The Stars My Destination*

---

[4] I'll return, later in this chapter, to the ethics of terraforming.

makes the powerful substance PyrE freely available to the crowds that he teleports to. The outcome is not revealed.

Isaac Asimov's *The End of Eternity* portrays meddlings in history by Eternity, a godlike agency that holds the secret of time travel. Each of Eternity's interventions stands to erase entire cultures, replacing them with something very different. It removes some people from history: either they never exist at all or they are replaced by analogues of themselves. This aspect of Eternity's work carries a taint even within the organization. Thus the Technicians—experts in making the actual changes—are somewhat ostracized even though their work is all-important to Eternity's goals.[5]

Cixin Liu's Remembrance of Earth's Past trilogy presents us with a variety of characters who seek—or who otherwise find themselves burdened with—great responsibility at moments when the fate of humanity is on their shoulders. Liu shows us the choices made by, among others, the bitter (yet naively hopeful) Ye Wenjie, the enigmatic Luo Ji, and the loving Cheng Xin. Another who makes critical decisions is Chu Yan, the deep-thinking, tactically brilliant captain of the spaceship *Blue Space*, whose choices are most unlike Cheng Xin's loving ones. The maverick Thomas Wade is largely thwarted by Cheng Xin in his efforts to take on supreme responsibility to protect humanity's future, but his efforts to press ahead with one technological project are highly consequential and his proposed solutions are never shown to be incorrect.

A common trope in science fiction is that a single person or small group can—whether through foresight and planning, improvization, or sheer inadvertence—alter the future course of humanity or, in many cases, that of an alien civilization. This often involves destroying a system of oppression that was in operation when he, she, or they arrived on the scene. Such plots are common in serial adventures—the long-running BBC series *Doctor Who* (1963–1989, 2005–present) is just one TV series that uses the familiar trope—but they also appear in science fiction novels with significant literary claims. Various operatives in Iain M. Banks's Culture series of novels and short stories make choices, and take actions, that entail great consequences. In *The Player of Games* (1988), for example, the sadistic Empire of Azad collapses from within when the Culture's greatest master of game strategies, Jernau Gurgeh, enters a tournament that allocates its public offices. Banks subverts many expectations, and Gurgeh is himself manipulated by the Culture's agents; still, this is a recognizable use of the trope.

---

[5] In the end, as I discuss in Chapter 4, we learn that Eternity did not understand how its careful, incremental acts to protect humanity were actually undermining it.

In its most debased form, the sense of wonder can be little more than a pornography of size and power, as when we're invited to gasp at the immensity of spaceships and other objects, or at the power of super-science weaponry. The latter includes powerful equipment for hand-to-hand combat, such as the lightsabers shown in the Star Wars movies, and extends to the galaxy-busting attack on the villainous Chlorians in E.E. "Doc" Smith's *Skylark DuQuesne*. As I discussed in Chapter 5, the ostensible theme of Smith's fiction is scientific responsibility, and particularly the responsible uses of powerful technology. However, the Skylark and Lensman series invite excitement over constant depictions of awe-inspiring, "incredible" (a favorite word), and very often destructive feats. We are led to identify with the heroes, and sometimes even the villains, as they perform their—well, yes—incredible actions. The most impressive villains, such as Marc DuQuesne, attract our sneaking admiration for their displays of power, though also for their courage and cunning.

As this plays out, Smith's novels engage overtly with questions about the responsible uses of science-based power. But they also provide enjoyable *fantasies* of vast power. The scale of these fantasies must surely be part of Smith's ongoing appeal for adolescent male readers. At the same time, it may be a reason why many other readers cannot take Smith, or perhaps the entire SF genre, seriously. I don't mean to sound puritanical about this. I enjoy these stories myself, even at their silliest, and I doubt that anyone is morally corrupted by them. It's all harmless. Still, such narratives eschew appeals to cognition in favor of power fantasies and spectacle.

The Trouble with Terrans (and Terraforming)

The sense of wonder can be debased, but it is salutary to reflect on humanity's place—and humanity's possible future—in the universe. Science fiction can open our eyes to levels of grandeur conveyed with imagination and a degree of intellectual care. Sometimes part of the point is to depict sublime beauty that is threatened by philistines and other wrongdoers.

James Blish's *A Case of Conscience* (discussed at some length in Chapter 3) can be read on one level as a story about the destruction of an idyllic planet and its civilization through human greed and power. At this level, we can factor out Father Ruiz-Sanchez's theological interpretation of Lithia as a trap set by Satan; we can even ignore Ruiz-Sanchez's story entirely, since he is ineffectual in preventing Paul Cleaver's plans to use Lithia for thermonuclear research. If we view *A Case of Conscience* in this way, Cleaver is the agent who drives events, and it is his catastrophic mistake that ultimately annihilates the entire planet. The story is his tragedy, since his arrogance and hubris destroy him along with Lithia.

This seems a legitimate, though incomplete, way to read Blish's novel. On this approach, the serpent in Lithia's garden turns out to be Cleaver—and humankind more generally, with its unwanted interference—rather than the so-called Snakes. At this level, Blish's novel resembles other such narratives about rapacious human colonists and their dire effects on alien planets, among them Ursula K. Le Guin's short novel *The Word for World is Forest* (1972), the James Cameron movie *Avatar*, and Phillip Mann's *The Disestablishment of Paradise: A Novel in Five Parts plus Documents* (2013).

In *The Word for World is Forest*, the imperialist and environmentally destructive actions of humans are presented very nearly as an allegory of Western colonialism and military ventures. The novel shows the fate of human colonists—a relatively small military group—who have set out to tame a planet that they call "New Tahiti" or, more officially, "Athshe." In the process, they enslave its intelligent inhabitants, the Athsheans, for whom they use the deeply derogatory term "creechies." Some colonists treat the diminutive Athsheans brutally, killing and raping them at will.

The colonists' mission is to log Athshe's vast, dark forests, shipping the lumber back to Earth and replacing the forests with grain fields. Driven to despair, however, the Athsheans rebel: they massacre a logging settlement with two hundred humans, then follow up with more, equally murderous, raids. They take special care to kill all women in the human camps to prevent future human births. From the Athsheans' viewpoint, the "yumens" are callous, physically enormous, heavily armed, and morally insane alien invaders. Some chapters are narrated from the viewpoint of Selver, the leader of the Athsheans' resistance, allowing us to see the invasion of their planet from the Athshean perspective, and to understand why the Athsheans respond with such intensity.

The parallel between the militarized humans and American military personnel during the Vietnam war is obvious, and it is scarcely surprising given the date of original publication in the early 1970s. Le Guin rubs in the lesson when one of the main human characters, Colonel Dongh, draws a parallel with failed attempts by Western armies to control his own ancestral land, Indo-China, in the face of popular resistance and unrelenting guerrilla tactics. In its historical context, *The Word for World is Forest* is a condemnation of American foreign policy in the late 1960s and early 1970s. It is a nod to the Vietnamese guerrillas who steadily prevailed against US military might in a long, cruel, asymmetrical war. Yet the resonance goes beyond this. *The Word for World is Forest* calls not only for peace and mutual respect within our species, but also for peace with non-human nature. In that sense, it carries an anti-anthropocentric and posthumanist message.

More than three decades later, the Hollywood movie *Avatar* achieved extraordinary commercial success with a remarkably similar plot. Several sequels are currently planned. As in *The Word for World is Forest*, the aliens in *Avatar* are shown as victimized and noble, while the intruders from Earth suffer a kind of insanity. Their aim is not logging the forests, though all the action takes place on a forested world. Instead, the humans employed by the Resources Development Administration, or RDA, have set up a heavily defended base on Pandora, a moon that we see circling a gas giant planet, to mine for a valuable mineral that they call "unobtanium." As the narrative commences, they have located a rich deposit beneath the gigantic Hometree of one clan of the alien Na'vi.

Pandora's atmosphere is poisonous to humans, but they are able to explore the surface by remote-controlling Na'vi-like bodies that they have learned to engineer using a combination of human and Na'vi DNA. These are the "avatars" that provide the film's title. Thus, the main character, a former Marine called Jake Sully, spends most of the film with his consciousness effectively residing in a Na'vi/human hybrid body.

The Na'vi are a blue-skinned humanoid species who are much taller and stronger than humans. They show superbly graceful movements and have a slightly feline appearance, but they look close enough to the human norm to permit strong audience identification. As well as being charmingly catlike, they evoke Native Americans as portrayed in classic Western films and more generally in popular culture. The Na'vi use bows and arrows for hunting, but their technological capabilities are more advanced than this suggests. They can ride the Ikran—winged, dragonlike creatures—as well as the local equivalents of horses. Even more impressively, they bond with the Ikran for life through a neural interface, having apparently co-evolved with them. (The obvious comparison here is with Anne McAffrey's Dragonrider series of novels, beginning with *Dragonflight* (1968).)

Pandora has a visually glorious ecosystem, with intricate neural connections among its covering of trees. These connections add up to a world mind, a global superhuman intelligence. The planet's fauna do not have these physical connections, with exceptions such as when the Na'vi connect to their individual Ikran. When the planet is threatened, however, its wild animals appear to be controlled in some way by the world mind, and they join in revolt against intruders. The Na'vi have access to the world mind, which they interpret as the goddess Eywa. As we see, they can even access the world mind to transfer a human being's consciousness permanently from their original body to their avatar.

In *The Word for World is Forest*, the diminutive Athsheans conduct their own successful revolt against human invaders, provoked by repeated atrocities. In *Avatar*, by contrast, the revolt of the physically impressive Na'vi is led by a renegade human: by Jake Sully after he falls in love with beautiful Neytiri, a prominent woman among her Na'vi clan. In desperation to save Neytiri's people and the world's biosphere, he establishes himself (in his avatar form) as a messianic leader among the Na'vi to challenge the military might of the RDA, commanded by the fanatical and terrifying Colonel Quaritch. The result is an apocalyptic battle between good (Na'vi) and evil (human) forces, with a few humans choosing the side of good and the future of the Na'vi and Pandora itself at stake.

An even more recent variation is Phillip Mann's *The Disestablishment of Paradise*. This tells the story of humanity's expulsion from a secular version of Eden. Mann uses an identified first-person narrator who is, herself, situated outside the main action and struggling to understand the course of events. By and large, the technique is successful, since Olivia has a distinctive and appealing voice. Near the beginning, she provides a memorable description of how the planet Paradise received its name. It was bestowed by a young spaceship captain during a close and sensual, almost sexual and loving, first encounter with the planet: Captain Estelle waded naked into the sea, claiming to be Aphrodite and to be reclaiming Paradise. Thus, the first human beings on the planet experienced it as welcoming, bountiful, and idyllic.

This episode suggests the Judeo-Christian story of Adam, Eve, and the fruitful Garden of Eden, though it also contains an obvious pagan reference. In fact, the narrator suggests that Captain Estelle was recalling Botticelli's painting "The Birth of Venus" and that the name "Paradise" probably had no specific biblical connotations for her. But whatever the spaceship captain might have thought, said, and done, on a gorgeous first day on Paradise, the novel's events play out as a sequence of human wrongdoing that results in a fall from grace.

As we read on, we are gradually introduced to a bleak back story, a chain of events after the planet's discovery but prior to the action of the main narrative. Delightful as Paradise first appeared, things soon went downhill. The human colonists who followed in Captain Estelle's path encountered bizarre creatures biologically similar to plants, but often mobile and dangerous. Among these were the baffling, perhaps subtly intelligent, Michelangelo-Reapers; the awe-inspiring dinosaur-like Dendrons, with their two long tree trunks like necks; and the prolific, damaging Tattersall Weeds. The colonists responded with destructive effect, using their weapons and tools to drive the Reapers and Dendrons toward extinction. All this exertion and danger notwithstanding, the early human generations found Paradise a profitable home until things gradually changed. Over decades, the planet itself began to alter. It increasingly resisted human efforts to occupy its surface and exploit its natural resources.

Suspense is not the point of this novel, although mystery certainly is. Throughout, the narrator weaves in information about how she came to compose her account of events. It is her attempt to witness and document the "disestablishment" of the planet—the human colonists' choice to withdraw from it—and to ask us to dwell upon how and why it took place. Like *The Word for World is Forest*, *The Disestablishment of Paradise* has a wider significance, with echoes of ecological disaster on Earth. Unlike Paradise, of course, planet Earth cannot consciously fight back and make us unwelcome as we pollute its land, sky, and seas, alter its global climate, and generally degrade its biosphere.

Much of the emphasis throughout *The Disestablishment of Paradise* is on the sensory experience of Paradise itself: on its scents and tastes and colors, its pleasures and dangers, and its grandeur. The planet lies beyond human comprehension, though there are ample hints that it possesses an overall intelligence of a kind. Paradise is not merely distant from us in interstellar space: it is profoundly alien to human expectations.

While works such as *A Case of Conscience*, *The Word for World is Forest*, *Avatar*, and *The Disestablishment of Paradise* imply a critique of humans damaging the ecologies of other planets—and an indirect critique of environmental destruction here on Earth—there is also a large body of science fiction that employs the idea of terraforming less critically. This reflects an attitude that favors expansion into space and understands the universe as inexhaustible in resources for human use. The icon of terraforming links, therefore, to the ethic of destiny that has such a strong presence in the SF field and was dominant during the Campbell Golden Age.

Although science fiction sometimes presents human beings adapting themselves biologically to extraterrestrial environments, it is relatively rare to find novels in which the idea of terraforming is specifically questioned and rejected. Writing in 2000, Sylvia Kelso (40–41) pointed to Alison Sinclair's *Blueheart* as one such example. *Blueheart* features a space-faring society that has terraformed numerous worlds. Until the terraforming process is completed, "adaptives" (humans biologically engineered to live in the original or partly terraformed environment) rely on various technological supports. But one world that is slated for terraforming—the eponymous Blueheart—has a surface almost completely covered by water, beneath which a complex ecosystem has evolved. Terraforming would devastate this; it would extinguish many of Blueheart's life forms including the floating kelp forests that dominate its seas.

During the long period of study that always precedes the start of terraforming, it has proved possible for human adaptives to live without any external support. Over three centuries, the sea-swimming adaptives have built a culture of their own, capable of sustaining itself indefinitely, and they have little interest in the space-faring human society outside. The central conflict is between these adaptives, living in the ocean, and a group who still intend to terraform the planet.

When a planet already has a biosphere (and perhaps some kind of intelligent life), the decision to reject terraforming may seem relatively straightforward: destroying the local ecosystem and imposing something else can seem like vandalism or worse. The case against terraforming, or more generally against large-scale engineering efforts on a planetary level or beyond, seems weaker when the planet concerned is, as it were, merely a rock in space. All of this arises for debate in Kim Stanley Robinson's Mars trilogy—*Red Mars*, *Green Mars*, and *Blue Mars*—undoubtedly the most noteworthy saga of terraforming to date.

The story proceeds over generations, beginning with the launch in 2026 of a colonizing ship to Mars with a mixed crew mainly from Russia and the US. The crewmembers have very different attitudes to the question of whether it is justifiable to terraform Mars to make it habitable to humans. The "Red" position that stands opposed to terraforming is led by Ann Clayborne, while Saxifrage Russell leads supporters of the pro-terraforming "Green" position. The Red position is itself motivated by a mix of concerns. In part, the Reds show a deep reverence for the landscape itself, with its vast, dry canyons, volcanic calderas, and meteor craters. For Clayborne, the main point is to preserve the surface of Mars in its primal form so it can be studied scientifically. However, as terraforming progresses, she suffers grief that goes beyond a mere intellectual preference and sense of professional frustration.

At the same time as they attempt to settle this debate, the colonists have to sort out the political relationship between their new home and Earth. One faction seeks to develop entirely new Martian political arrangements, traditions, and *mores*, and to escape Earth's authority. Earth is itself increasingly dominated by transnational corporations that dwarf the power of traditional governments. In *Blue Mars*, the final novel of the trilogy, the factions on Mars hammer out a new constitution for their emerging, independent world.

Science Fictional Societies

Referring to stories about alien societies, the philosopher Bernard Williams states, disparagingly, "Since they can offer no concrete resistance at all to the most primitive fantasies, the results are pathetically or repulsively impoverished" (1985, 200). This is not entirely unfair. A reader looking for sharp insights into the anxieties, foibles, and character types in her own society will probably not find them in many science fiction stories. The exceptions are likely to be novels that are set in the present, or the very near future, and emphasize the characters' reactions to something new that disrupts the social order. If I am writing about a very different society, perhaps arising in the future or perhaps located in outer space or on another planet, a different set of considerations will come into play.

First, consider the challenge if I choose to write fiction about my own society. As I do so, my imagination will be restrained by facts about how my society actually works: among other things, by facts about its scientific knowledge base, available technologies, methods of production, mode of political organization, laws, *mores*, practices, institutions, and everyday habits of life. I might be extended some artistic license by readers and reviewers: for example, they'll allow me to focus on particular sub-cultures or to exaggerate for satirical effect. At the end of the day, however, I'll be held to high standards of verisimilitude. Likewise, if I write about societies well documented by anthropologists or historians then established facts about those societies will impose a certain discipline. In brief, I can't just make stuff up.

Once I start writing about distant planets, lost worlds on Earth, or future human societies, much of that restraint disappears. If I wish, I am free to imagine whatever social arrangements I need to provide a thin background for tales of adventure and conflict in exotic locations. There is no "concrete resistance," as Williams expresses it, from the reality of any particular society that has existed in the actual world. It does not follow, however, that anything goes or that science fiction writers have it easy. It is a considerable imaginative feat to conceive of—and then somehow convey to readers or viewers—societies that have never existed.

Some science fiction writers can get away with generic solutions, possibly involving quasi-medieval political arrangements combined with advanced forms of transport and weaponry. Depending on what else these narratives have to offer, their authors might deserve Williams' stricture. For other writers, however, or for the same writers on other occasions, there is a challenge in imagining more thickly realized societies. These might be extrapolations from our own society and its tendencies—as with much SF set in the relatively near future—or they might be societies radically discontinuous from our own, as with much SF set in the far future or on other planets. We are significantly handicapped in attempting to imagine what any of these sorts of societies might actually be like, just as people from earlier eras would have found it difficult to imagine twenty-first century societies with their greatly changed scientific understanding, technological capability, methods of production, political organization, laws, *mores*, institutions, practices, and everyday habits.

To make an imagined society appear plausible—intuitive to readers or viewers with their rich, if largely tacit, understandings of the world—and yet also convincingly strange and worth thinking about, requires special skill plus intensity in its application. When the task is done well, science fictional societies strike us as real, coherent, and "lived in"; we suspend our disbelief. More than that, these societies engage our moral imaginations: we find ourselves condemning or admiring them, or discovering that (and why) we cannot do either without reservation.

Any science fiction stories that extend beyond the briefest vignettes must also engage our sympathies with characters who are somewhat realistic. We need to care enough about their problems to continue reading or watching. The novels and stories in Banks's Culture series are a case in point. They take place in a setting that includes many human civilizations at various stages of social and technological development, many alien civilizations, likewise at various stages of development, and vast structures that include spaceships the size of present-day cities and worlds the size of solar systems. Yet these books tell the ongoing evolution of the Culture only indirectly. They are focused on the choices of vividly realized characters.

In *Matter*, for example, the stakes are high. For most of the novel, the conflict that drives the narrative relates to power politics in a barbaric and relatively low-tech civilization occupying the inner surface of one concentric sphere (later two) within a vast artificial "Shellworld." Throughout, however, we're given hints that even larger forces are in play. An ambitious alien species, the Oct, is manipulating the Shellworld's human civilizations for its own purposes, which ultimately endanger the entire Shellworld and its numerous civilizations—both human and alien. This background dynamic leads to a series of ultraviolent events (not unusually for Banks) as the story reaches its climax.

All of this intrigue and action is on a very large scale, but *Matter* is a novel of character and psychology. It shows the choices of a small group of characters, each of whom is placed in a series of unwanted and perilous situations. Three siblings—and the servant of one of them—rise to the challenges in different ways. They are psychologically transformed by their circumstances, their own responses, and the resources—internal and external—that they manage to draw upon at times of crisis. Though Banks describes enormous structures and momentous events, there is always a focus on characters and choices. In particular, we see the resourcefulness and courage developed by two royal brothers who had never required those qualities in their privileged lives.

Ann Leckie's Imperial Radch series provides another case in point. This series commenced in 2013 with *Ancillary Justice*, followed by *Ancillary Sword* (2014) and *Ancillary Mercy* (2015). Here, Leckie shows one way to keep readers' interest in extraordinary beings who operate in a space opera setting where much is at stake. She depicts military operatives known as "ancillaries": Artificial Intelligences downloaded into human bodies. The main character of the series, Breq, is the superintelligence that formerly controlled a spaceship called the *Justice of Toren*. While many science fiction narratives rely on the Pinocchio effect, depicting artificial conscious entities that yearn to be human, Leckie takes the opposite approach. Following the destruction of the *Justice of Toren*, Breq has been downloaded into a human body and experiences this as limiting compared to the much greater capabilities of a spaceship's controlling intelligence.

Throughout the series, Breq deals—Solomon-like—with large and small injustices. In the background are the intentions of a powerful and unpredictable alien species, the Presger, and a war within the powerful intelligence Anaander Mianaai, which has split in two after its vengeful destruction of an entire star system. The most delightful character in the series may be Translator Zaiat, an intelligence with a human body created by the Presger for the purpose of interacting with humans. Without any human socialization, Zaiat repeatedly draws what seem strange—though strangely logical—conclusions, and acts in bizarre ways by the standards of any human culture. Among other things, Zaiat develops an obsession with the taste of fish sauce.

One fragment of Anaander Mianaai has endorsed the great injustice that it committed in the past, while the other fragment now regards its actions with regret. The first Anaander fragment is callous and power hungry, while the second shows kindness and responsibility for the wellbeing of others. Breq aligns with this gentler fragment of Anaander, while solving various dilemmas through wisdom and diplomacy as much as force.

The Presger only reluctantly acknowledge humans as possessing what they call Significance—worthiness of moral consideration. As the story reaches its climax in *Ancillary Mercy*, the question is whether they are also prepared to recognize the Significance of human-made Artificial Intelligences. Breq's request that they do so is a roll of the dice during a final confrontation with the "bad" version of Anaander, and the result has potentially enormous ramifications for a human civilization that relies on AIs as its servants. The Presger don't determine the issue, but merely by taking it up for consideration they force an end to the conflict. Thus Breq becomes a liberator to the AIs and—at least for the moment, and likely for longer—a successful rebel against the existing order of dominance and subordination.

Concluding Remarks

To an outsider, science fiction might seem no more than a lurid form of entertainment aimed at children and adolescents, and dominated by visually spectacular adventure stories. Any moral content lies at the level of injunctions to use our powers for good, coupled, perhaps, with assurances that good will prevail over evil if only we show enough resourcefulness and courage. In all honesty, much science fiction takes exactly this form, but it is not the entirety of the genre. Science fiction is an important cultural response to the scientific and industrial revolutions of the past four centuries, and over the past *two* of those centuries its exponents have established a rich, varied, self-interrogating tradition.

From the seventeenth century onward, developments in science have led to new understandings of the universe, the future, and ourselves. Our own biological species, *Homo sapiens*, stands revealed as the outcome of evolutionary pressures over millions of years. Anthropocentrism and human exceptionalism are challenged from numerous directions. All human societies are evidently mutable, and since the Industrial Revolution of the eighteenth and nineteenth centuries it has become natural to imagine future societies quite different from our own. Increasingly, in the twentieth and twenty-first centuries, we have seen new technologies that function as more than tools. That is, they alter the functioning of our bodies rather than merely extending our bodies. In the future, it's a safe bet, technology will be used for purposes such as lengthening the maximum span of life and augmenting human cognition.

These developments create opportunities for storytelling, and science fiction writers have embraced them. They speculate about radical technological innovations and about the future of humanity itself. They set adventures and conflicts in prehistoric environments, or in distant space, or in one or another imaginary future. They posit new kinds of minds: non-human, or dubiously human, Intelligent Others. These include extraterrestrial aliens, remarkably advanced robots and AIs, and mutated humans. Aliens, robots, and mutants can seem frightening, but they have an allure. Sympathy for the

Intelligent Other may be nudging science fiction writers toward a posthumanist, post-anthropocentric ethic. This may stand as an alternative to, or perhaps form a hybrid with, an ethic of human destiny.

Science fiction portrays efforts to enhance human powers of action. It offers maps of what might await round the next bend of history—within our own lifetimes, maybe, or those of our children. It can warn, entice, or excite us with portrayals of technocratic hells, Arcadias of artificial abundance, or amoral cybertopias that appear more and more to be reflected in our real-world cities. As I hope to have shown, even science fiction stories set in the distant future, or at vast distances in space, can speak to our condition in myriad ways. The genre's greatly extended narrative possibilities enable storytellers to question traditional social arrangements and moral norms—and offer alternatives—and to dramatize moral dilemmas.

I hope to have made some contribution to informed science fiction *criticism*. If they want to be relevant when they comment on science fiction, literary and cultural critics must engage with the genre's over-arching mega-text. Though SF's various icons and tropes do not have single, fixed meanings, they attract clusters of available meanings, and they direct attention to particular questions, themes, ideas, and attitudes. Understanding science fiction requires sufficiently deep familiarity with its canon and traditions to read its mega-text intuitively.

Science fiction is not for everyone. It is highly adaptable and varied, but it has thematic concerns of its own. For some people, no doubt, scientifically informed speculations about the future, our nature as a species, and our place in a universe of vast space and deep time ⋯ simply don't appear interesting. Science fiction will leave those people cold. For me—and perhaps for you, if you've read this far—it's entirely different. Those same speculations seem urgent and important. Where better to go than science fiction to see them fleshed out in narrative?

References

Blackford, R. (2013). The great transition: Ideas and anxieties. In M. More & N. Vita-More (Eds.), *The transhumanist reader: Classical and contemporary essays on the science, technology, and philosophy of the human future* (pp. 421–429). Chichester, West Sussex: Wiley-Blackwell.

Blackford, R. (2014). *Humanity enhanced: Genetic choice and the challenge for liberal democracies.* Cambridge, MA: MIT Press.

Hayles, N. K. (1999). *How we became posthuman: Virtual bodies in cybernetics, literature, and informatics*. Chicago and London: University of Chicago Press.

Kelso, S. (2000). Tales of earth: Terraforming in recent women's SF. *Foundation: The International Review of Science Fiction, 78*, 34–43.

Williams, B. (1985). *Ethics and the limits of philosophy*. London: Fontana.

Wolfe, C. (2010). *What is posthumanism?* Minneapolis and London: University of Minnesota Press.

Chronological List of Works Discussed

This is a selected list, limited to those novels, movies, and other creative works that I have discussed more than in passing. Unfortunately, there is no ideal way to set out such a list. I have settled on chronological order, except that I have grouped items by the same author or in the same movie franchise. When in doubt, I have erred on the side of including inherently important works if they receive more than a mention in the text.

The number of works listed for a particular author is not necessarily an indication of the relative space given to discussion of the author in the main text. For example, I list only two works for Samuel R. Delany, but one of them, *Stars in My Pocket Like Grains of Sand*, is discussed at substantial length.

Although the works I have chosen for discussion have been strongly influenced by my particular focus on science fiction and moral philosophy, what follows could function as a reasonable introductory reading and viewing list for people wishing to immerse themselves in the SF genre.

Early Modernity

Johannes Kepler. *Somnium, Sive Astronomia Lunaris* (completed c. 1608–1609, but not formally published until 1634).

Jonathan Swift. *Gulliver's Travels* (1726; full original title *Travels into Several Remote Nations of the World. In Four Parts. By Lemuel Gulliver, First a Surgeon, and then a Captain of several Ships*).

© Springer International Publishing AG 2017
R. Blackford, *Science Fiction and the Moral Imagination*, Science and Fiction,
DOI 10.1007/978-3-319-61685-8

From Shelley to Wells

Mary Shelley. *Frankenstein; or, The Modern Prometheus* (1818).
———. *The Last Man* (1826).
Edgar Allan Poe. "Mellonta Tauta" (1848).
Jules Verne. *Twenty Thousand Leagues under the Sea* (first serialized 1869–1870).
Robert Louis Stevenson. *Strange Case of Dr Jekyll and Mr Hyde* (1886).
Edward Bellamy. *Looking Backward, 2000–1887* (1888).
H.G. Wells. *The Time Machine* (1895).
———. *The Island of Doctor Moreau* (1896).
———. *The Invisible Man* (1897).
———. *The War of the Worlds* (1897).
———. *When the Sleeper Wakes* (first serialized 1898–1899; revised and published in book form as *The Sleeper Awakes* 1910).
———. *The First Men in the Moon* (1901).
———. *The Food of the Gods and How it came to Earth* (1904).
———. *The World Set Free* (1914).

Twentieth-Century Developments

E.M. Forster. "The Machine Stops" (1909).
Edgar Rice Burroughs. *A Princess of Mars* (1912).
Charlotte Perkins Gilman. *Herland* (1915).
Karel Čapek *R.U.R.* [play] (1920).
———. *The Makropulos Affair* [play] (1922).
———. *War with the Newts* (1936).
George Bernard Shaw. *Back to Methuselah (A Metabiological Pentateuch)* [play sequence] (published 1921; first produced 1922).
Metropolis (dir. Fritz Lang; novelization of the screenplay 1926, cinematic release 1927).
E.E. "Doc" Smith. *The Skylark of Space* (1928).
———. *Galactic Patrol* (1937).
———. *Gray Lensman* (1939).
———. *Skylark DuQuesne* (1966).
Olaf Stapledon. *Last and First Men* (1930).
———. *Odd John: A Story Between Jest and Earnest* (1935).
———. *Sirius* (1944).
Philip Wylie. *Gladiator* (1930).

———. *The Disappearance* (1951).

Aldous Huxley. *Brave New World* (1932).

———. *After Many a Summer* (1939; also known as *After Many a Summer Dies the Swan*).

———. *Ape and Essence* (1948).

———. *Island* (1962).

King Kong (dir. Merian C. Cooper and Ernest B. Schoedsack, 1933).

Things to Come (dir. William Cameron Menzies, 1936).

C.S. Lewis. *Out of the Silent Planet* (1938).

Don A. Stuart (John W. Campbell, Jr.). "Who Goes There?" (1938).

A.E. van Vogt. *Slan* (1940).

———. *The Weapon Makers* (first serialized 1943; published in book form 1947 and in revised form 1952).

Robert A. Heinlein. *Methuselah's Children* (first serialized 1941; expanded and published in book form 1958).

———. *Tunnel in the Sky* (1955).

———. *Starship Troopers* (1959; originally published in serial form as *Starship Soldier*).

———. *Stranger in a Strange Land* (1961).

———. *The Moon is a Harsh Mistress* (1966).

———. *Friday* (1982).

Arthur C. Clarke. *Against the Fall of Night* (first serialized 1948; expanded and published in book form 1953).

———. *The Sands of Mars* (1951).

———. *Childhood's End* (1953).

———. *The City and the Stars* (1956).

———. *2001: A Space Odyssey* [novel] (1968).

———. *Rendezvous with Rama* (1973).

George Orwell (Eric Arthur Blair). *Nineteen Eighty-Four* (1949).

Isaac Asimov. "The Evitable Conflict" (1950).

———. *Foundation* (1951).

———. *Foundation and Empire* (1952).

———. *Second Foundation* (1953).

———. *The End of Eternity* (1955).

———. *Foundation's Edge* (1982).

———. *The Robots of Dawn* (1983).

———. *Robots and Empire* (1985).

Ray Bradbury. *The Martian Chronicles* (1950; subsequently published in the UK in slightly different form as *The Silver Locusts*).

———. *Fahrenheit 451* (1953).

Philip José Farmer. "The Lovers" (1952).
——. *The Lovers* (1961).
Vercors (Jean Bruller). *Les Animaux Dénaturés* (1952).
Theodore Sturgeon. *More than Human* (1953).
——. *Venus Plus X* (1960).
William Golding. *Lord of the Flies* (1954).
John Wyndham (full name John Wyndham Parkes Lucas Beynon Harris). *The Chrysalids* (1955; published in the US as *Re-Birth*).
James Blish. *The Seedling Stars* (1956).
——. *A Case of Conscience* (1958).
Forbidden Planet (dir. Fred M. Wilcox, 1956).
Alfred Bester. *The Stars My Destination* (first serialized 1956–1957; first published in book form 1957 as *Tiger! Tiger!*).
Kurt Vonnegut. *The Sirens of Titan* (1959).
——. *Cat's Cradle* (1963).
Walter M. Miller. *A Canticle for Leibowitz* (1960).
Stanislaw Lem. *Solaris* (1961).
Philip K. Dick. *The Man in the High Castle* (1962).
——. *Do Androids Dream of Electric Sheep?* (1968).
Frank Herbert. *Dune* (1965).
Star Trek [television series] (1966–1969).
2001: A Space Odyssey [film] (dir. Stanley Kubrick, 1968).
John Brunner. *Stand on Zanzibar* (1968).
Ursula K. Le Guin. *The Left Hand of Darkness* (1969).
——. *The Word for World is Forest* (1972).
——. "The Ones Who Walk Away from Omelas" (1973).
——. *The Dispossessed: An Ambiguous Utopia* (1974).
Arkady and Boris Strugatsky. *Roadside Picnic* (1972).
Joe Haldeman. *The Forever War* (1974).
Joanna Russ. *The Female Man* (1975).
Samuel R. Delany. *Trouble on Triton: An Ambiguous Heterotopia* (1976; originally published as *Triton: An Ambiguous Heterotopia*).
——. *Stars in My Pocket like Grains of Sand* (1984).
Marge Piercy. *Woman on the Edge of Time* (1976).
——. *He, She and It* (1991; subsequently published in the UK as *Body of Glass*).
Frederik Pohl. *Man Plus* (1976).
James Tiptree Jr. (Alice B. Sheldon) "Houston, Houston, Do You Read?" (1976).
Gregory Benford. *In the Ocean of Night* (1977).

——. *Across the Sea of Suns* (1984).

——. *Great Sky River* (1987).

——. *Tides of Light* (1989).

——. *Furious Gulf* (1994).

——. *Sailing Bright Eternity* (1995).

——. *The Martian Race* (1999).

——. *The Sunborn* (2005).

Star Wars (dir. George Lucas, 1977; now known as *Star Wars: Episode IV—A New Hope*).

Rogue One: A Star Wars Story (dir. Gareth Edwards, 2016).

Octavia E. Butler. *Kindred* (1979).

Barry B. Longyear. "Enemy Mine" (1979).

Gene Wolfe. *The Shadow of the Torturer* (1980).

Vernor Vinge. "True Names" (1981).

——. *Marooned in Realtime* (1986).

Blade Runner (dir. Ridley Scott, 1982).

Damien Broderick. *The Judas Mandala* (1982; revised in an Australian edition 1990).

——. *Transcension* (2002).

Connie Willis. "Fire Watch" (1982)

——. *Doomsday Book* (1992).

William Gibson. *Neuromancer* (1984).

The Terminator (dir. James Cameron, 1984).

Margaret Atwood. *The Handmaid's Tale* (1985).

Orson Scott Card. *Ender's Game* (1985).

Watchmen [comic-book limited series] (scripted Alan Moore, 1986–1987).

Iain M. Banks. *Consider Phlebas* (1987).

——. *The Player of Games* (1988).

——. *Use of Weapons* (1990).

——. *Matter* (2008).

Michael Crichton. *Jurassic Park* [novel] (1990).

Greg Egan. *Quarantine* (1992).

——. *Permutation City* (1994).

Jurassic Park [film] (dir. Steven Spielberg, 1993).

Jurassic World (dir. Colin Trevorrow, 2015).

Nancy Kress. *Beggars in Spain* (1993).

——. *Beggars and Choosers* (1994).

——. *Beggars Ride* (1996).

Kim Stanley Robinson. *Red Mars* (1993).

——. *Green Mars* (1994).

——. *Blue Mars* (1996).
——. *New York 2140* (2017).
Mary Doria Russell. *The Sparrow* (1996).
——. *Children of God* (1998).
Alison Sinclair. *Blueheart* (1996).
Michael Marshall Smith. *Spares* (1996)
Melissa Scott. *Dreaming Metal* (1997).
Ted Chiang. "Story of Your Life" (1998).
Mike Resnick. *Kirinyaga: A Fable of Utopia* (1998).
The Matrix (dir. The Wachowskis, 1999).

Twenty-first Century

A.I. Artificial Intelligence (dir. Steven Spielberg, 2001).
Cixin Liu (Liu Cixin). *The Three-Body Problem* (2006; trans. Ken Liu 2014).
——. *The Dark Forest* (2008; trans. Joel Martinsen 2015).
——. *Death's End* (2010; trans. Ken Liu 2016).
X-Men: The Last Stand (dir. Brett Ratner, 2006).
Avatar (dir. James Cameron, 2009)
Watchmen [film] (dir. Zack Snyder, 2009).
Her (dir. Spike Jonze, 2013).
Ann Leckie. *Ancillary Justice* (2013).
——. *Ancillary Sword* (2014).
——. *Ancillary Mercy* (2015).
Phillip Mann. *The Disestablishment of Paradise: A Novel in Five Parts plus Documents* (2013).
Transcendence (dir. Wally Pfister, 2014).
Ex Machina (dir. Alex Garland, 2015).
Passengers (dir. Morten Tyldum, 2016).

Index

© Springer International Publishing AG 2017   **201**
R. Blackford, *Science Fiction and the Moral Imagination*, Science and Fiction,
DOI 10.1007/978-3-319-61685-8

CPSIA information can be obtained
at www.ICGtesting.com
Printed in the USA
LVOW13s0014120917
548250LV00041B/1787/P

9 783319 616834